SOTHIS PRESS

Sothis Press aims to completely reimagine the public domain book with an emphasis on universal access to captivatingly designed print editions. We believe that technology will enable every single book ever published to be perpetually accessible in beautiful print form. Sothis is beauty in ubiquity.

www.sothis.press
1 Waterhouse Square, London, ECIN 2ST

Star Lore of All Ages

A Collection of Myths, Legends, and Facts, Concerning the Constellations of the Northern Hemisphere

William Tyler Olcott

SOTHIS PRESS

ISBN 978-1-0365-0830-2

Star Lore of All Ages is
in the public domain.

This edition is published by
Sothis Press © 2023. The company
is registered in England and Wales.
The Company Number is 14900389.

Star Lore of All Ages

A Collection of Myths, Legends, and Facts Concerning the Constellations of the Northern Hemisphere

By

William Tyler Olcott

Author of "A Field Book of the Stars," "In Starland, with a Three-Inch Telescope," etc.

With 50 Illustrations in the Text and 64 Full-Page Illustrations

G. P. Putnam's Sons
New York and London
The Knickerbocker Press
1911

To

G. H. O.

AND

L. S. S.

PREFATORY NOTE

THE author's first book, *A Field Book of the Stars* was simply intended as a guide to the constellations. It was an effort on his part to acquaint the reader with the star groups and the individual star names. In his book, *In Starland with a Three-inch telescope* he sought to indicate to the amateur astronomer what could be seen of the stellar wonders with a modest telescopic equipment.

It follows naturally that having come to be on friendly terms with the stars, and having seen many of the beautiful sights that the night reveals, the tyro should wish to know more of the history of the stars and how the constellations came to be named, and the purpose of this book therefore is to satisfy that desire.

It is always a pleasure to trace back to their sources the traditions with which time has endowed the enduring, and thus the study of the myths and legends that surround the eternal stars possesses a surpassing charm for those who have learned to know them intimately and through nightly communion with them have come to love them.

The author quotes extensively from R. H. Allen's *Star Names and Their Meanings,* an exhaustive and scholarly work and an authority on the subject, and he here pays tribute to the author for the pleasure a close perusal of his book affords, and heartily commends it to all those who desire to make a closer study of the philology of the ancient star names.

INTRODUCTION

THERE are many persons who are familiar with the bright stars and constellations of these northern latitudes who are unaware of the beautiful myths and legends that time and fancy have woven about them.

As even a meagre knowledge of star lore has added greatly to the writer's pleasure in the study of the stars, and has served to render their appearance full of suggestion, he has been interested in collecting for this volume a portion of that varied history of the heavens that has been presented in terms imaginative by the peoples of all ages. Those who admire the beauty of the stars may learn to love them by reason of the literary and legendary associations recalled by their appearance.

Much that appears in these pages has been published from time to time in books on popular astronomy of comparatively recent date, but to the writer's knowledge no comprehensive story has as yet been presented of the constellations, and of the stars they contain.

In the compilation of this volume, the purpose has been to include all matter pertinent to the subject, in order that the history of the constellations, as known and as written by all nations in every age, might be arranged in convenient form for the benefit of those who only know the stars by sight.

A further aim has been to revive an interest in the mythology that twines about the stars. It has seemed but right that this wealth of star lore, buried deep in the treasury of the past, should once more see the light, and add its increased charm and interest to those who scan the skies.

Such a history must ever serve to keep bright the memory

of the earliest times, and fanciful though the constellation figures seem, our stars bear the same names that were given to them in the very dawn of civilisation.

In conclusion, it is hoped that the history of the heavens here set forth will awaken fresh interest in the stars, and will secure for them the attention that is their just due, on the part of all lovers of the beautiful.

W. T. O.

NORWICH, CONN, January, 1911.

And all the signs through which Night whirls her car,
From belted Orion back to Orion and his dauntless Hound,
And all Poseidon's, all high Zeus's stars,
Bear on their beams true messages to man.

Poste's Translation of Aratos.

CONTENTS

ILLUSTRATIONS

BIBLIOGRAPHY

In the compilation of this volume the author hereby acknowledges his indebtedness to the following publications for much valuable information.

STAR-NAMES AND THEIR MEANINGS	Richard H. Allen
STARLAND	Sir Robert S. Ball
INFLUENCE OF THE STARS	Rosa Baughan
A HISTORY OF ASTRONOMY	Arthur Berry
ASTRONOMICAL MYTHS	J. F. Blake
STELLAR THEOLOGY PRIMITIVE CONSTELLATIONS	Robert Brown, Jun.
GEOGRAPHY OF THE HEAVENS	Elijah H. Burritt
THE STORY OF THE STARS	George F. Chambers
THE SYSTEM OF THE STARS	Agnes M. Clerke
THE SIDEREAL HEAVENS	Thomas Dick
STAR LORE	J. A. Farrer
METRICAL PIECES	N. L. Frothingham
ASTRONOMICAL ESSAYS	James E. Gore
HOW TO KNOW THE HEAVENS	Eward Irving
ASTRONOMY OF THE ANCIENTS	Sir George C. Lewis
DAWN OF ASTRONOMY STAR-GAZING	Sir Joseph Norman Lockyer
THE FRIENDLY STARS	Martha Evans Martin
THE ASTRONOMY OF THE BIBLE	E. W. Maunder
THE CHILDREN'S BOOK OF STARS	Geraldine E. Mitton
THE STARS	Simon Newcomb
ASTRONOMY OF PARADISE LOST	T. N. Orchard
FAMILIAR TALKS ON ASTRONOMY	W. H. Parker
HISTORY OF THE HEAVENS	The Abbé Pluche
ANCIENT CALENDARS AND CONSTELLATIONS	E. M. Plunket
THE STARS IN SONG AND LEGEND	J. G. Porter
THE STORYLAND OF STARS	Mara L. Pratt
STORIES OF STARLAND	Mary Proctor
MYTHS AND MARVELS OF ASTRONOMY THE FLOWERS OF THE SKY THE EXPANSE OF HEAVEN	Richard A. Proctor

HANDBOOK OF THE STARS	N. J. Rolfe
ASTRONOMY OF THE OLD TESTAMENT	Giovanni Schiaparelli
CURIOSITIES OF THE SKY	Garret P. Serviss
PLEASURES OF THE TELESCOPE	Garret P. Serviss
ASTRONOMY WITH AN OPERA-GLASS	Garret P. Serviss
ASTRONOMY WITH THE NAKED EYE	Garret P. Serviss
NEW ASTRONOMY	David P. Todd
CELESTIAL OBJECTS FOR COMMON TELESCOPES	Rev. T. W. Webb
HISTORY OF THE INDUCTIVE SCIENCES	William Whewell
ORIENTAL AND LINGUISTIC STUDIES	William D. Whitney

JOURNAL OF AMERICAN FOLK-LORE
AMERICAN ORIENTAL SOCIETY'S JOURNAL
MEMOIRS OF THE LONDON ANTHROPOLOGICAL SOCIETY
POPULAR ASTRONOMY
THE WORKS OF JOHN PLAYFAIR

The Origin and History of the Ancient Star Groups

THE ORIGIN AND HISTORY OF THE ANCIENT STAR GROUPS

> Some man of yore
> A nomenclature thought of and devised,
> And forms sufficient found.
> . . . . . .
> So thought he good to make the stellar groups,
> That each by other lying orderly,
> They might display their forms. And thus the stars
> At once took names and rise familiar now.
>
> ARATOS.

THE origin of the constellations is still open to conjecture, for, though all nations since the dawn of history have recognised these ancient stellar configurations, and at one period or another employed them in some symbolic or representative capacity, the fact remains that the researches of archæologists have failed to yield definite proof as to who first designed them and where they were first known.

There is little doubt that the constellations were the result of a deliberate plan, as La Place affirms. Possibly they were an endeavour on the part of some patriarch of the ancient world to grave an imperishable record of a great event, or a series of noteworthy occurrences in the world's history, for all posterity to read, and although no Rosetta stone has been found as yet to enable the present race of man to decipher their meaning, still the problem attacked by the ablest savants of all nations has yielded theories respecting the origin and purposes of the constellations that cannot be far from the truth.

In the very dawn of the world, when human instinct first inspired observation, primitive man began to look about him and take stock of his environment. The daily wants of nature supplied, the natural phenomena would claim man's attention, and first he would take cognisance of the sun, moon, and stars that provided life's chief essential, light.

For purposes of identification alone, there must have been at an early date certain designations for the individual stars that gave rise to all subsequent stellar nomenclature. The sun, moon, and planets, the brighter luminaries, would first excite man's interest and attention, and then the brightest stars would attract and mystify him.

As time went on, observation would soon indicate to human intelligence the relationship of the sun and moon to the fixed stars, and the seasonable difference in the appearance of the nocturnal skies.

All this would be in strict accord with the natural laws of the observational faculties. Such elementary knowledge of the heavenly bodies would presently lead to the establishment of certain facts relative to the stars, features concerning their apparent change in position, that if marked would render a service to the race.

Very early in the history of the world the stars must have served to record the passage of time, a service they have faithfully and accurately rendered mankind through all the ages to the present day.

The first tillers of the soil must have marked well the stars, and certain of them doubtless proclaimed the time of sowing and reaping. The circumpolar stars guided the rude crafts of the early navigators, and unquestionably in the earliest times they singled out "the star that never moves," Polaris, as an unfailing and reliable beacon to direct their course.

The rising and setting of the stars thus became matters of paramount importance, governing alike the actions of the husbandmen and those who sailed the seas. Certain stars were also indicative of impending meteorological changes, and their appearance at particular seasons was watched for with keenest interest.

The wonder and mystery the stars inspired, and their

utility in daily life, soon led to their becoming objects of idolatry, and as their importance increased, astrology, that pseudo-science, Kepler's "foolish daughter of a wise mother," sprang into being, and for a time suppressed, discouraged, and hampered the legitimate and scientific study of the heavens.

Thus early in the history of man we find the stars all-important to his welfare. No course was pursued or plan adopted without first consulting the heavenly bodies. They governed alike the policies of nations and the actions of individuals. They ruled absolutely over the destinies of the high and lowly, the rich and poor, and horoscopes became a necessity of life, and divination the highest pursuit of man.

In Sabianism, or star worship, we have, therefore, the earliest form of religion, and in astrology and the adoration of the stars the progenitors of the modern science of astronomy.

From this universal attention to the stars, there sprang up the myriad fancies and peculiar notions, the products of imagination, that peopled the sky with animals and quaint figures, and gave rise to the constellated stellar groups that have come down to us, and figure on the modern charts of the heavens.

There are many traditions that have emerged from the mists that shroud the distant past respecting the origin of the constellations, and the science of astronomy, and as that origin is antediluvian, the knowledge that we have of the subject must perforce be largely traditional in its character.

An early tradition affirms that the immediate descendants of Adam cultivated a knowledge of the stars, and that Seth and Enoch inscribed upon two pillars, one of brick, the other of stone, the names, meanings, secret virtues, and science of the stars, with the divisions of the zodiac.

Josephus states that he saw in Syria the pillar of stone, which alone remained in his day. The history of two mysterious pillars entwined with a serpent, the symbol of revolution, can be traced through all the ages, from remote antiquity until it reaches our dollar sign

Then there is a tradition that has survived the ages, that Noah, who was also known as Oannes and Janus, was the inventor of astronomy. It is certain that Noah and his family were soon worshipped and inextricably mixed with stars and gods.

The Chaldeans attributed their knowledge of the stars to Noah, who became a two-faced deity, as he could look backwards and forwards. He was known as "the God of Gates," as he opened the door which God shut, and Noah and the Ark became Janus and Jana, solar and lunar deities. Of all this tradition meets us everywhere.

It is a remarkable fact that, from the earliest times, as far as we can judge from the cuneiform inscriptions and hieroglyphics that have been deciphered, the sign for God was a star.

Astronomy unites with history and archæology in pointing to the Euphrates Valley, and, as we might expect, the region of Mt. Ararat, as the home of those who originated the ancient constellation figures.

Authorities agree, for the most part, that the originators of Sabianism and stellar lore in this region were not the Semitic Babylonians, but a people generally termed "Akkadians," a word meaning highlanders, or mountaineers, the most ancient race known to us, who came down from the mountainous region of Elam or Susiana, to the east of Assyria, bringing with them the rudiments of writing and civilisation.

The Babylonians, previous to the invasion of the Akkadai, unquestionably had some knowledge of the stars. It was thought in those early times that the mountains on the east supported the firmament, and that the zenith was fixed over Elam. There were observatories established in all the large cities of Chaldea, many of the shrines on the topmost terraces being dedicated to this purpose, and at an early date the stars were named and numbered.

The Babylonian Tablets, the oldest records extant, reveal that the Akkadians introduced their sphere and zodiac into Babylonia before the year 3000 B.C., and the zodiac of the Akkadians corresponds almost exactly with the signs we know to-day.

It seems almost folly to endeavour to set the date of the invention of the constellations, for that period must approximate the age of the habitable world, and in all probability the stellar figures known to us were not designed at any one time, and lost their originality by the varying conditions that time has wrought in the past, for even in comparatively recent years there have been many attempts to alter them.

Bailly, a brilliant scholar and eminent astronomer, contends that the phenomena of astronomy had been closely observed before the great races of mankind separated from the parent stock. He claims, and few would dispute him, an antediluvian race as the originators of astronomical science. In proof of this he cites the fact that there are ancient Persian records which refer to the four famous "Royal Stars" as having marked the four colures (the meridian points of the solstices and equinoxes), a fact only possible in antediluvian times.

Maunder, who has made a very careful study of archæology in its relation to the constellational figures, has revealed many interesting features in connection with them. He writes:

"The first feature which the old constellation figures present to us is a very striking one. They cover only a portion of the heavens, and a large region roughly circular in the southern hemisphere is left entirely vacant. Swartz was the first to make the significant suggestion that this space was left vacant because the inventors of the constellations lived too far north to permit of their viewing this part of the heavens."

Pursuing this line of thought, Maunder considers that the designers of the figures lived, in all probability, between 36° and 42° north latitude, so that the constellations did not originate in Egypt or Babylon. By computing where the centre of the vacant space coincided with the southern pole, we get the date 2800 B.C., which was probably the date when the ancient work of constellation making was completed.

It has been remarked that among the constellation figures conspicuous by their absence are the following animals: the elephant, the camel, the hippopotamus, the crocodile, and the tiger, so it is reasonably safe to assume that neither India, Arabia, nor Egypt was the birthplace of the sphere. Greece, Italy, and Spain may be excluded on the ground that the lion figures as one of the constellations. We have left Asia Minor and Armenia, a region bounded by the Black, Mediterranean, Caspian, and Ægean seas, as the logical birthplace of the stellar figures. The fact that we find a ship among the stars warrants us in believing that it is on the coast of this country, and not in its interior, that we should expect to find the land where the constellations were first known.

The division of the zodiac into twelve signs, the number of months in the year, is one of very great significance, for we infer from the fact that it was so arranged to assist in the observation of the position of the sun among the stars.

Many of the authorities hold that the zodiac was planned while the spring equinox fell in the constellation Taurus. In support of this claim it may be said that, if this is the case, the sun was ascending all through the signs that face the east, and was descending all through the signs that face the west, a significant and logical arrangement which could hardly be accidental.

The date of the zodiac is given as 3000 B.C., which agrees very well with the significant position of the four Royal Stars previously mentioned which marked the four cardinal points, and were thus especially prominent.

A close inspection of the stellar groups yields many points of interest, notably the fact that everywhere there is indication of design and not chance in the arrangement and configuration. There seems to have been a definite idea in some one's mind respecting them, a desire to perpetuate a vitally important record. It may be of interest to mention a few of the facts that have inclined scholars to this belief:

To begin with, we find many figures duplicated, and in most cases the two figures are close together in the sky. Thus we see the figures of two Dogs, two Bears, two Giants subduing Serpents, each pair in close proximity. Then there are two Goats, two Crowns, two Streams, and two Fishes bound together.

The zodiacal constellations are often clearly connected with neighbouring figures. We observe the Bull attacked by the Giant Hunter Orion, Aquarius pouring a stream of water into the mouth of the Southern Fish, the Scorpion attempting to sting Ophiuchus, and the Ram pressing down the head of the Sea Monster.

Again, one portion of the sky was known to the ancients as "the Sea," and here we find, as we might expect, many marine creatures,—the Dolphin, the Whale, the Fishes, the Sea Goat, and the Southern Fish.

Other features in support of the theory of design are found in the half-figures, Pegasus, Taurus, and Argo, and the so-called Deluge group, comprising the Ship stranded on a rock, the Bird, the Altar, the Centaur offering a sacrifice, and the Bow set in the Cloud.

It is supposed that, at a time far remote, the Akkadians were conquered by the Semitic race, and that the conquerors imposed only their language on the conquered, adopting, it is said, the Akkadian mythology, laws, literature, and system of astronomy.

At an early date in the world's history we find astronomy and astrology flourishing in China, India, Arabia, and Egypt.

The early astronomical annals of the Chinese reveal the fact that, before the year 2357 B.C., the Emperor Yao had divided the twelve zodiacal signs by the twenty-eight mansions of the moon.[1]

The Arabians are said to have received their astronomical knowledge from India, and in China, Arabia, and India we find an almost identical system, *i.e.*, that of the Lunar Stations, or Lunar Mansions, employed to indicate the daily progress of the moon amid the stars.

India has been claimed as the birthplace of the constellation figures, but modern research, says Allen, finds little in Sanscrit literature to confirm this belief.

There is a controversy as to whether Indian astronomy was derived from Greece or independent of it. In support of the latter theory, it is said that the Brahmins were too proud to borrow their science from the Greeks or Arabs, and also that it was improbable that two rival Hindu sects, the Brahmins and Buddhists, should have adopted the same innovations in their calendars and religious symbolism. Again, the Greeks held Indian astronomy in high esteem, while the Hindus only bestowed a moderate praise on the Grecian science.

The Egyptians, on whose early monuments the twelve zodiacal signs are found, acknowledged that they derived their knowledge of the stars from the Chaldeans, and they were in turn the teachers of the Greeks as early as the time of Thales and Pythagoras.

Herodotus states that the Egyptians were the first of all mankind who invented the year and divided it into twelve parts, a statement much at variance to the accepted testimony of the Babylonian Tablets.

Of the constellations outside the zodiac, we find a few groups and stars mentioned at an early date, notably in the Old Testament, where, in the Book of Job, there are references to the Bear, Orion, and the Pleiades, names that have come down to us. Homer and Hesiod both mentioned

[1] It is of interest to note that the Chinese were called "Celestials" because their empire was divided after the Celestial spaces.

the same constellations, which is indicative of the importance of these star groups in the eyes of the ancients. Hesiod also refers to the stars Arcturus and Sirius, and these two stars may well be considered the most ancient of all the stars from the standpoint of stellar nomenclature.

Authorities differ as to the source from which the Greek knowledge of the stars was derived, but in all probability it did not come from any one source but was imported from Egypt, Chaldea, and Phœnicia.

The founder of the science of astronomy in Greece was Thales, the head of the Ionic School of Philosophy, a citizen of Miletus, who lived about 540 B.C. It is said that he first taught the Greek navigators to steer by the Little instead of the Great Bear.

Eudoxus, a native of Cnidus, who lived about the fourth century B.C., a contemporary of Plato, was the first Greek who described the constellations with approximate completeness. He is reported to have visited Egypt and to have there received astronomical instruction. He wrote *The Enoption*, or *The Mirror*, and *The Phenomena* or *Appearances*, both prose works and unfortunately not extant, but Aratos, the Alexandrine poet, versified the latter work about 270 B.C., and it has descended to our day.

Aratos was a native of Soli in Cilicia, and Court Physician to Antigonus Gonatas, King of Macedonia. He was a contemporary of Aristophanes, Aristarchus, and Theocritus, and he always mentions the constellations as of unknown antiquity. His sphere accurately represented the heavens of about 2000 B.C. His poem has been considered an authority on stellar nomenclature, and has been closely followed by all subsequent delineators of the constellation figures.

This sphere of Eudoxus, which has been transmitted to us through the verses of Aratos, contained forty-five constellations, twenty in the northern hemisphere, twelve in the southern, and thirteen in the zodiacal group, the

Pleiades being considered as a separate constellation in addition to Taurus.

Allen makes the following interesting reference to this famous poem: "When the poem entitled *The Phenomena* of Aratos was introduced at Rome by Cicero and other leading characters, we read that it became the polite amusement of the Roman ladies to work the celestial forms in gold and silver on the most costly hangings, and this had previously been done at Athens, where concave ceilings were also emblazoned with the heavenly figures."

The Phenomena is the most ancient description of the constellations extant, and has been translated into all languages. Cicero and Germanicus Cæsar both made translations of it, and no less than thirty-five Greek commentaries on the work are known to us.

Eudoxus considered the heavens as divided up into constellations with recognised names. "He did not deal with the stars singly, but gave a sort of geographic description of their territorial position and limits, according to groups, distinguished by a common name." His work's chief value consists in the comprehensive view of the heavens it affords, and in the description of the constellated heavens in their entirety.

Although the contributions of Eudoxus and Aratos to astronomical literature are highly regarded and authoritative, the acknowledged founder of our scientific astronomy is Hipparchus, who was the first to discover the perpetual and apparent shifting of the stars known as the Precession of the Equinoxes. Only two of his works have come down to us, his *Commentary*, and the reproduction of his Star Catalogue by Ptolemy, who was known as "the Prince of Astronomers." This catalogue enumerated 1022 stars, of which 914 form constellations, and 108 are unformed. It is held in much respect even by modern astronomers, and agrees in the main with the enumeration of Aratos. Procyon, however, appears as a constellation, and the asterism Equuleus, the foremost Horse, is added,

Ptolemy
National Museum, Naples

an asterism that figures on modern star maps. The observations of Hipparchus were made between 162 and 127 B.C., while those of Ptolemy embodied in the *Syntaxis*, as his work was entitled, were made from 127 to 151 A.D.

The *Syntaxis* was practically an epitome of the results of the early star-gazers of Greece and Western Asia, and comprised a list of 1028 stars classified in forty-eight constellations. Each star is named by its position in the figure supposed to include the stars of the group. Thus the constellation Draco contains thirty-one stars, some of which received the following descriptive names: "the star upon the tongue," "the star in the mouth," "the star above the eye," etc. This method of naming the stars continued in use until the eighteenth century, when a letter or a number with the Latin genitive of the constellation was used. In Ptolemy's catalogue appears the first comparative list of stellar magnitudes.

The constellations of the Greeks were ultimately accepted and adopted by the Persians, Hindus, Arabs, the nations of Western Asia, and the Romans, from whom they have been borrowed by the modern world. To Greece, then, we are indebted for the figures now depicted on our celestial globes and the many interesting myths associated with them, notably the legend of Perseus and Andromeda, which is fully illustrated in the starry skies.

Although the savages of prehistoric times first bequeathed the stellar configurations to science, we listen to their harsh ideas, as Bacon puts it, "as they come to us blown softly through the flutes of the Grecians."

From the time of Ptolemy till the year 1252, no advance of importance was made in the matter of cataloguing the stars, but in this latter year there appeared the celebrated Alphonsine Tables compiled by Arabian or Moorish astronomers at Toledo under the auspices of the subsequent King Alphonso X., known as "the Wise."

A correction of Ptolemy's sphere was published by the Arabian astronomer Ulugh Beg in 1420 A.D., in which there was a description of the constellations derived from Al-Sufi's translation of five centuries previously.

The catalogues of Copernicus and Tycho Brahe followed, the former's great work laying the foundations of modern astronomy. In 1603 the *Uranometria* of Johann Bayer appeared in Germany. This chart contained forty-eight constellations and a list of 709 stars. Bayer invented the system in vogue to-day of denoting each star by a letter of the Greek alphabet, the brightest star in each figure being designated Alpha with the Latin genitive of the constellation. It was soon found that the stars in many of the groups exceeded the number of letters in the alphabet, and such stars were denoted by the letters of the Roman alphabet.

Succeeding Bayer's catalogue there appeared consecutively the charts of Bartsch, Schiller, Kepler, Royer, Halley, and in 1690 that of Hevelius, who added the asterisms of the Hunting Dogs, the Giraffe, the Lizard, the Unicorn, the Lynx, the Sextant, Fox and Goose, and Sobieski's Shield, all recognised by modern astronomers.[1]

Flamsteed's catalogue, published in 1719, comprised fifty-four constellation figures, and exhibited a new method of stellar designation, the stars being consecutively numbered in the order of their right ascension, a method employed in modern charts for the fainter stars.

La Caille, known as "the true Columbus of the southern sky," in his publications of 1752 and 1763, invented fourteen new star groups which included the names of many instruments of the sciences and fine arts, the majority of which have been rejected by modern delineators of the constellations.

[1] In the case of the charts of Bartsch and Schiller it is of interest to note that these astronomers endeavoured to do away for all time with the old constellation names, and Christianise, so to speak, the stellar hosts. On their charts the twelve Apostles were each represented by a constellation, and other Biblical names were substituted for the time-honoured figures. It is needless to add that this nomenclature was not popular and failed of general adoption.

Subsequently Le Monnier, Bode, and Lalande published stellar catalogues, adding new asterisms, the latter's chart containing a total of eighty-eight constellations.

In 1840 the famous German astronomer Argelander published his star catalogue, the most complete that had appeared up to that time. It contained 210,000 stars. Argelander brought order where there had been much confusion, by separating one constellation from another by irregular boundary lines, so that all the stars would be embraced within the borders of some stellar figure. His system is employed in many of the modern charts of the heavens.[1]

To-day there are over a hundred large catalogues of the stars, but there is a discrepancy in the number of constellations accepted by astronomers. Prof. Young recognised sixty-seven as in ordinary use, and in these northern latitudes about fifty-five are generally known.

Allen tells us that "eighty or ninety may be considered as now more or less acknowledged, while probably a million stars are laid down on the various modern maps, and this is soon to be increased perhaps to forty million on the completion of the present photographic work for this object by the international association of eighteen observatories engaged upon it in different parts of the world."

In conclusion, it may be of interest to review briefly the conception of the firmament in vogue in ancient times among the different nations of the old world.

The Persians are said to have considered 3000 years ago that the whole heavens were divided up into four great districts, each watched over by one of the "Royal Stars," Aldebaran, Antares, Regulus, and Fomalhaut.

[1] Photography has played an important part in stellar catalogues of recent years, Kapteyn's chart made up from plates taken at Cape Town containing over 300,000 stars, and every year approximately 2000 plates of the heavens are taken by the astronomers in charge of the Harvard College Observatory Station at Arequipa, Peru.

The Assyrians looked upon the stars as divinities, endowed with beneficent or evil powers.

Among the Chaldeans the sky was regarded as a boat, shaped like a basket. The space below was the earth, which was flat and surrounded by water.

The Egyptians worshipped Osiris and Isis as ancestors, and showed Plutarch their graves, and the stars into which they had been metamorphosed.

The ancient Peruvians thought that there was not a beast or bird on earth whose shape or image did not shine in the sky. They considered the luminaries and stars guardian divinities and worshipped them. They also thought that the stars were the children of the sun and moon.

The Hebrews had a notion that the sun, moon, and stars danced before Adam in Paradise.

The Bushmen, or early inhabitants of Africa, regarded the more conspicuous stars as men, lions, tortoises, etc. They believed that the sun, moon, and stars were once mortals on earth, or even animals, or inorganic substances which happened to get translated to the skies.

In New Zealand heroes were thought to become stars of greater or less brightness according to the number of their victims slain in battle.

The North American Indians believed that many of the stars were living creatures, and knew Ursa Major as a Bear, the same figure known in the Far East.

The Tannese Islanders divided the heavens into constellations with definite traditions to account for the canoes, ducks, and children that they see in the skies.

In the South Pacific islands dying men will announce their intention of becoming a star, and even mention the particular part of the heavens where they are to be looked for.

The Eskimos thought that some of the stars had been men and others different sorts of animals and fishes, which was also the mythical belief of the Greeks and Romans.

According to Slavonic mythology the stars are regarded as living in habitual intercourse with men and their affairs.

An ancient legend was that there were no stars till the giants of old, throwing stones at the sun, pierced holes in the sky, and let the light of that orb shine through the holes which we call stars,—and Anaximenes thought that the stars were fixed in the dome of heaven like nails.

Thus we find, as some one has put it, that "astronomy like a golden thread runs through history and binds together all tribes and peoples of the earth," and the girdle of stars we view nightly remains as the most ancient monument of the work of intelligent man, "the oldest picture book of all."

Andromeda
The Chained Lady

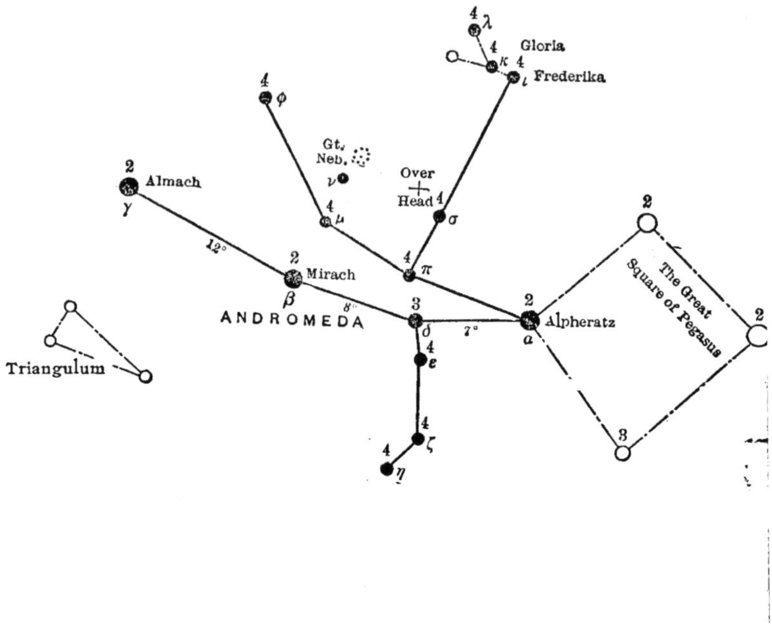
4 λ
4 κ
Gloria
4 ι
Frederika
4 φ
Gt. Neb.
ν
Over Head
4 σ
4 μ
2 Almach
γ
12°
2 Mirach
β
4 π
ANDROMEDA
8°
3 δ
7°
2 Alpheratz
α
2
The Great Square of Pegasus
2
Triangulum
4 ε
4 ζ
4 η
3

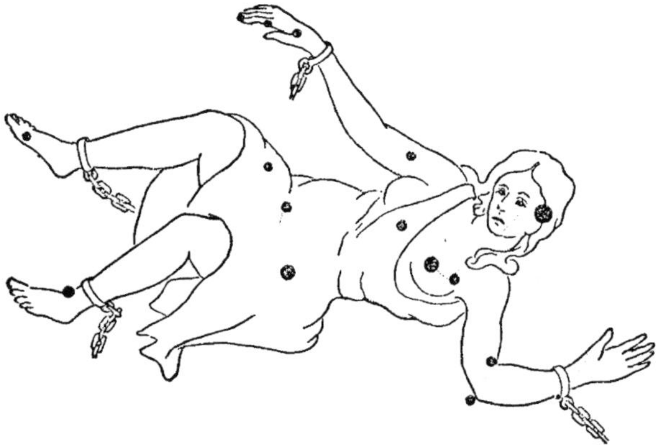

ANDROMEDA

ANDROMEDA
THE CHAINED LADY

And there revolves herself, image of woe,
Andromeda, beneath her mother shining.

ARATOS.

THE origin of the constellation known to us as Andromeda is lost in remote antiquity, but the myth that relates to Andromeda, the daughter of Cepheus and Cassiopeia, and associated with the constellation, is probably as well known to-day as any that has come down to us. According to this myth, Cassiopeia boasted that she was fairer than the sea nymphs. This attitude was offensive to Neptune, who despatched a monster of the deep to ravage the seacoast. Cassiopeia, terrified at the prospect, besought the aid of the all-powerful Zeus, who ruled that her daughter Andromeda must be sacrificed to appease the wrath of the sea god. Consequently Andromeda, amid great lamentation, was chained to a wave-washed rock, there to await the coming of the sea monster to devour her.

In accordance with this legend, we find the constellation Andromeda depicted in the old star atlases as a beautiful maiden chained to a rock, with Cetus the Whale or the sea monster represented near at hand about to devour her.

In Burritt's atlas,[1] Andromeda is represented with chains attached to her wrists and ankles. The rock to which she was said to have been bound does not appear in the picture.

[1] *Geography of the Heavens*, by Elijah H. Burritt.

In the edition of the Alphonsine tables, Allen tells us Andromeda is pictured with an unfastened chain around her body, and two fishes, one on her bosom and the other at her feet, showing an early connection with the neighbouring constellation Pisces.

In the Leyden Manuscript, Andromeda is represented as lying partly clothed on the sea beach, chained to rocks on either side, and on a map printed at Venice in 1488 she is pictured as bound by the wrists between two trees.

The legend further relates that Perseus, flying through the air on his steed Pegasus, fresh from his triumph over the Medusa, espied the maiden in distress, and like a true champion flew to her assistance.

> Chained to a rock she stood; young Perseus stay'd
> His rapid flight, to woo the beauteous maid.

Holding the Medusa's head before him, he assailed the sea monster that threatened Andromeda, and immediately the creature was turned to stone, and the hero had the pleasure of releasing the wretched maiden.

For the statement that Perseus when he freed Andromeda was mounted on his winged steed Pegasus, there is however no classical authority.

The constellation Andromeda is bounded on the west by Pegasus, and on the east by Perseus, and thus links the two constellations together. This doubtless accounts for the presence of Pegasus in the myth.

Brown[1] thinks that in this legend of Andromeda and Perseus we have but another version of the all-pervading solar myth. Perseus may be Bar-Sav, the solar Herakles, and Andromeda his bride Schachar (the morning red).

The Hindus have almost the same story in their astronomical mythology, and almost the same names that have come down to us. They call the constellation "Antarmada." In an ancient Sanscrit work are found draw-

[1] *Stellar Mythology*, by Robert Brown, Jr.

Photo by Hanfstaengl

Perseus and Andromeda

ings of Antarmada chained to a rock with a fish beside her.

Sappho, the Greek poetess of the 7th century B.C., refers to Andromeda, and Euripides and Sophocles both wrote dramas about her,—but there is little doubt, as Allen states, that the constellation originated far back of classical times in the valley of the Euphrates.

Plunket[1] is of the opinion that the constellation of Andromeda dates from 3500 B.C. in accordance with the other constellations around it, and there is some ground for believing that its date goes back to 6000 B.C.

In Dr. Seiss's mythology, Andromeda was intended for a prophetic symbol of the Christian church. Sayce claims that she appeared in the great Babylonian Epic of Creation of more than two millenniums before our era, in connection with the story of Bel Marduk and the dragon Tiamat, which doubtless is the foundation of the story of Perseus and Andromeda.

The constellation Andromeda has borne the following names:

Mulier Catenata, the woman chained.

Persea, as the bride of Perseus.

Cepheis, from her father.

Alamac, from the title of the star Gamma.

Some authorities claim that Andromeda was a native of Æthiopia and regard her as a negress. The Arabian astronomers knew these stars as "Al mar'ah al musalsalah," and to them they represented a sea calf or seal with a chain around its neck that united it to one of the two fishes.

Allen states that according to Cæsius, Andromeda represented the biblical Abigail of the Books of Samuel, and Julius Schiller in 1627 made of these stars the Sepulchrum Christi, the new Sepulchre wherein was never man yet laid.

[1] *Ancient Calendars and Constellations*, by E. M. Plunket.

Milton in his *Paradise Lost* thus refers to Andromeda:

> the fleecy star that bears
> Andromeda far off Atlantic seas
> Beyond the horizon.

Kingsley's *Andromeda* is beautifully descriptive of the constellation.

Pluche[1] accounts for the names of the constellations Perseus, Andromeda, and Cepheus in the following ingenius way:

It was an ordinary turn of the Hebrew and Phœnician languages to say that a city or country was the daughter of the rocks, deserts, rivers, or mountains that surrounded her or that were enclosed within her walls. Thus Jerusalem is often called "the daughter of Sion," that is, the daughter of drought or daughter of the barren hills contained within its compass. Palestine originally was nothing more than a long maritime coast consisting of rocks and a sandy flat shore. It was proper to speak of this long coast as the daughter of Cepheus and Cassiope, Cepha signifying a stone. If you would say in Phœnician, a long coast or a long chain or ridge, you would call it Andromeda. Palestine would have been destroyed had it not been for the assistance of the barks and pilots that voyaged to Pharos and Sais to convey provisions. Strabo informs us that the Phœnicians were accustomed to paint the figure of a horse upon the stern of their barks, but there was beside the winged horse (the emblem of navigation) a horseman bearing a peculiar symbol, and, as it were, the arms of the city of Sais. This was the Medusa's head. Furthermore, a bark in the vulgar tongue was called Perseus, which means a runner or horseman. This then according to Pluche was the meaning of the fabled sacrifice of Andromeda:—Exposed to a cruel monster on the rocks of Joppa, in Syria, Andromeda (or the coast towns of Palestine), owed her deliverance to a flying rider, Perseus

[1] *History of the Heavens*, by Abbé Pluche

(the Phœnician barks), to whom the goddess of Sais had lent the frightful head of Medusa to turn all her enemies into stone with terror. Josephus wrote that in his day the inhabitants of Joppa showed the links and remains of the chain that bound Andromeda to the rock, and the bones of the sea monster.

Burritt suggests that the fable of Andromeda might mean that the maiden was courted by some monster of a sea captain who attempted to carry her away, but was prevented by another more gallant and successful rival.

Maunder[1] claims that in the 12th chapter of the Apocalypse there is an allusion to what cannot be doubted are the constellations Andromeda, Cetus, and Eridanus: "And the serpent cast out of his mouth after the woman, water as a river, that he might cause her to be carried away by the stream." Andromeda is always represented as a woman in distress, and the sea monster has always been understood to be her persecutor, and from his mouth pours forth the stream Eridanus.

The constellation Andromeda presents a beautiful appearance rising in the eastern sky in the early evening during the months of autumn. Low over the hills twinkle her chain of stars, sweeping down in a long graceful curve from the Great Square of Pegasus, like tiny lamps swinging from an invisible wire, a chain of gold with which heroic Perseus holds in check his winged steed.

Astronomically speaking, the great feature of interest in the constellation is the famous nebula, the so-called "Queen of the Nebulæ," or Al Sufi's "Little Cloud," said to have been known as far back as A.D. 905. In the West it seems to have been first observed by Simon Marius, Dec. 15, 1612. It is the only naked eye nebula, and according to Marius it resembles "the diluted light from the flame of a candle seen through horn." An arc light glimpsed through a dense fog is also descriptive of its

[1] *The Astronomy of the Bible*, by E. M. Maunder.

naked eye appearance.[1] It is an enormous body, estimated to be in length as much as thirty thousand times the distance of the earth from the sun (ninety-three million miles), a proportion inconceivable. Herschel thought that the nebula was resolvable into separate stars, although his glass failed to prove the fact. Later observations with more powerful telescopes confirmed his opinion. An examination made at Cambridge in 1848 proved the existence of upwards of fifteen hundred minute stars within the nebula, while the nebulous character of the whole was still apparent. In the spectroscope this nebula gives clearly a continuous spectrum, thus proving that it is not a mass of incandescent gas but rather a highly condensed cluster of stars. Recent and more reliable calculations of its distance give it a light journey of about nineteen years.

The star Alpha Andromedæ, or Alpheratz as it was called by the Arabs, was formerly associated with the constellation Pegasus, and called Delta Pegasi. The Arabs also knew this star as "Sirrah," and it represented to them the horse's navel. Alpheratz is situated at the north-eastern corner of the Great Square of Pegasus, a stellar landmark, and is known as one of the "Three Guides," marking the equinoctial colure, the prime meridian of the heavens, Beta Cassiopeiæ and Gamma Pegasi being the other two guides. In astrology Alpheratz portended honour and riches to all born under its influence. It culminates at 9 P.M., on the 10th of November. Alpheratz is situated in the head of the figure of Andromeda, and was familiarly known as "Andromeda's Head" in England two centuries ago. In all late Arabian astronomy taken from Ptolemy it was described as the "Head of the Woman in Chains." According to Prof. Russell, Alpheratz has a dark companion spectroscopically revealed, revolving about it in a

[1] While Serviss says it resembles a whirlwind of snow, and the appearance of swift motion and terrific force is startling.

Great Nebula in Andromeda

highly eccentric orbit, in a period of about one hundred days.

Gamma Andromedæ was known to the Arabs as "Almach." Allen tells us this name was derived from a phrase meaning a small predatory animal similar to a badger. The propriety of such a designation here is not obvious in connection with Andromeda, and the name would indicate that it belonged to a very early Arab astronomy. In the astronomy of China, Gamma, with other stars in Andromeda and Triangulum, was "Tien Ta Tseang," "Heaven's Great General." Astrologically this star was "honourable and eminent." The duplicity of Almach according to Allen was discovered by Johann Tobias Mayer of Göttingen in 1778, and Wilhelm Struve in 1842 found that its companion was a close double. Herschel regarded Almach as one of the most beautiful objects in the heavens, and Webb, Proctor, and Serviss all speak in glowing terms of the beautiful contrast in colour between the gold and blue of the primary and its companion. Almach certainly vies in beauty with the famous double Beta Cygni, and is perhaps with this exception the most charming of all double stars. It is an easy double for small telescopes and is consequently a great favourite with amateur astronomers. It requires a 5″ glass at least to split the blue companion star. The celebrated meteor shower known as "the Andromedes II.," the so-called Bielid meteors of November, radiate from the vicinity of this star. There was a wonderful display of these meteors in 1872 and 1885. Delta Andromedæ marks the radiant point of the Andromedes I., a meteor shower due the 21st of July.

The fourth magnitude stars λ, κ, ι Andromedæ and the fifth magnitude star ψ Andromedæ form a "Y"-shaped figure which bears the name of "Gloria Frederica" or Frederick's Glory, an asterism formed by Bode in 1787 in honour of the great Frederick II., of Prussia, who died in 1786. The figure is thus described: "Below a nimbus the sign of royal dignity hangs, wreathed with the imperishable laurel of fame, a sword, pen, and an olive branch, to distinguish this ever to be remembered monarch, as hero, sage, and peacemaker." This figure, with the exception of the nimbus, appears on Burritt's Atlas, but later atlases omit the asterism entirely, and it is seldom mentioned.

The remaining stars in this constellation require no special mention.

Aquarius
The Water Bearer

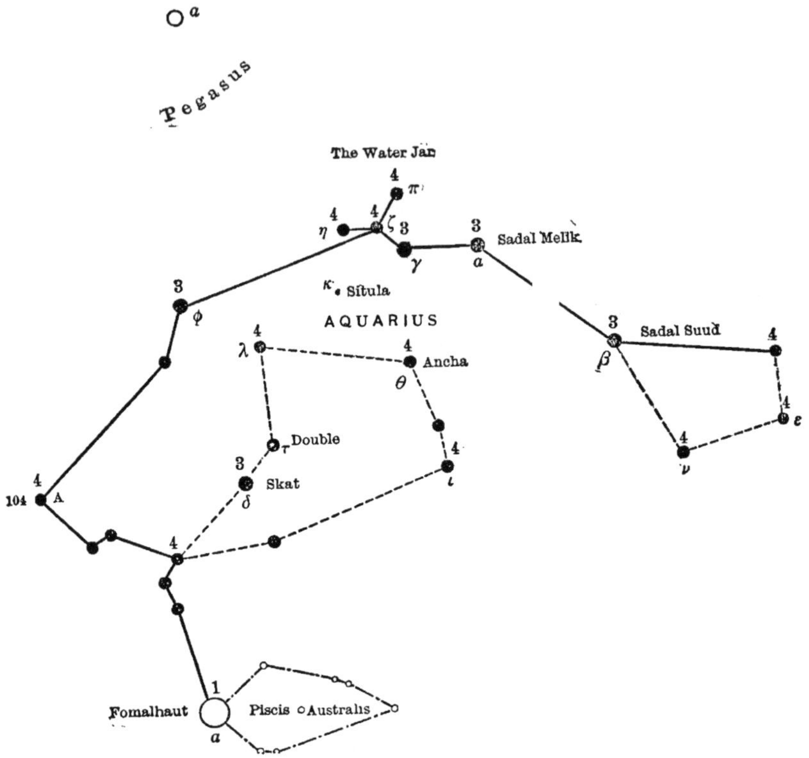
Pegasus
The Water Jar
Sadal Melik
Situla
AQUARIUS
Sadal Suud
Ancha
Double
Skat
104
Fomalhaut
Piscis Australis

AQUARIUS

AQUARIUS
THE WATER BEARER

While by the Horse's head the Water-Pourer
Spreads his right hand.

ARATOS.

THE astronomers of all nations, with the exception of the Arabians, have adopted the figure of a man pouring water from a jar or pitcher to express this constellation. The Arabs, being forbidden by law to draw the human figure, have represented this sign by a saddled mule carrying on his back two barrels of water, and sometimes by only a water bucket. They called the constellation "Al-Dawl," the "Well Bucket," and not the "Water Bearer."

For some reason, all the ancients imagined that the part of the sky occupied by the Water Bearer and neighbouring constellations contained a great celestial sea. Here we find the Whale, the Fishes, the Dolphin, the Southern Fish, the Sea Goat, the Crane, (a wading bird), and even Eridanus, the River Po, is sometimes shown as having its source in the Waterman's Bucket. It also seems appropriate that Pegasus is situated in this region of the sky, for the winged horse was the Phœnician emblem of navigation, and the star Markab, as Alpha Pegasi was called by the Arabs, signifies a ship or vehicle.

According to Ideler, the reason for this designation of "the Sea" for this region of the heavens is because the sun passes through this part of the sky during the rainy season of the year.

An Egyptian legend averred that the floods of the Nile were caused by the Water Bearer sinking his huge urn into the fountains of the river to refill it, and accordingly this constellation represented to the Egyptians the rainy period of the winter season. However, the Egyptians were probably indebted to some other people for their knowledge of this constellation, for Egypt is not a land subject to heavy rains.

Aquarius is represented even on very early Babylonian stones as a man or boy pouring water from a bucket or urn; around the waist is a scarf, part of which is held up by the left hand. For some reason, which is lost to us, his right arm is stretched backward to the fullest extent possible so as to reach over almost the entire length of the constellation Capricornus, which bounds Aquarius on the west.

The significance of the pouring of the water from the urn into the mouth of the Southern Fish is also unaccounted for. The conception is such a singular and striking one that it was evidently the result of design rather than fancy. Maunder referring to this peculiar figure says: "Strangely enough through all the long centuries that the starry symbols have come down to us, Aquarius has always been shown as pouring forth his stream of water into the mouth of a fish, surely the strangest and most bizarre of symbols."

According to Norse mythology, Aquarius was considered Wali's palace, and it was supposed to be covered with silver. In the Indian zodiac, the name of the constellation is "Kumbha," meaning "Water Jar." Allen states that Kumbha is from κομβη, or Storm-god. Here again we find the constellation associated with rain and tempest.

Brown tells us that Aquarius in the Hebrew zodiac represented the tribe of Reuben, "unstable as water."

In Greek mythology, Aquarius represented Ganymede, the cup-bearer of the gods. Ganymede was a beautiful youth of Phrygia, and the son of Tros, King of Troy. He was taken up to heaven by Jupiter as he was tending his father's flocks on Mt. Ida, and became the cup-bearer of the gods in place of Hebe.

Photo by Anderson

Ganymede and the Eagle
Museum of Vatican, Rome

In a Roman zodiac, Aquarius was represented by a peacock, the symbol of Juno, the Greek Herē, in whose month Gamelion (Jan.-Feb.) the sun was in this sign. Aquarius has also been represented as a goose, another bird sacred to the goddess.

In February, the Aquarius month the sun entered the Peruvian sign known by the name "Mother of Waters" and "Eagle Bridge." The Water Mother was figured as a sacred lake located in the Southern Fish and the Crane. The month of February marks the height of the rainy season in the Andes, and the rivers are then in flood so that the powers of the Mother of Waters are at this season most conspicuously displayed.

Allen[1] states that the New Testament Christians of the 16th and 17th centuries appropriately likened Aquarius to John the Baptist and to Judas Thaddæus the Apostle. In Babylonia this constellation was associated with the 11th month (Jan.-Feb.), called "Shabatu," meaning "the Curse of Rain," and the Epic of Creation has an account of the Deluge in its 11th book, corresponding to this the 11th constellation, each of its other books numerically coinciding with the other zodiacal signs. In that country an urn seems to have been known as "Gu," meaning a water-jar overflowing. Plunket tells us that "Gu" is possibly an abbreviation of "Gula," the name of a goddess. This goddess under another name was a personification of the dark water or chaos, hence the identification of the goddess Gula with the constellation Aquarius.

In the cuneiform inscriptions of western Asia we read: "The planet Jupiter in the asterism of the Urn lingers." Considering the imagined aqueous nature of this region of the sky it is not difficult, as Plunket says, to understand how the Vedic Rishis, who appear to have combined the characteristics of poets, scientists, and observers of the heavens, should have in 3000 B.C., when the sun was in

[1] *Star Names and Their Meanings*, by Richard H. Allen.

conjunction with Aquarius at the time of the winter solstice, have described the fire of the solstitial sun as "hiding in, being born in, and rising out of the celestial waters of the constellation Aquarius."

Some suppose Aquarius represents Deucalion, who was placed among the stars after the celebrated deluge of Thessaly in 1500 B.C., and the creation legend connected with this constellation identifies it with the Flood. It may be that Noah, desiring to perpetuate the record of the Deluge, found in the scroll of night a parchment that never fades, and in the stars characters that time cannot efface.

Aquarius has also been identified with Cecrops, the Egyptian who journeyed to Greece and founded Athens.

Proctor in his *Myths and Marvels of Astronomy* tells us that Aquarius astrologically speaking is in the house of Saturn. Its natives, those born between Jan. 20th and Feb. 19th, are robust, steady, strong, and healthy, and of middle stature, delicate complexion, clear but not pale, sandy hair, hazel eyes, and generally of honest disposition. It governs the legs and ankles, and reigns over Arabia, Petræa, Tartary, Russia, Denmark, Lower Sweden, Westphalia, Hamburg, and Bremen. It is masculine and fortunate, and an aqueous blue colour is attributed to it.

The Anglo-Saxons called Aquarius "se Waeter-Gyt," the "Water Pourer," and it was also known by the queer title "Skinker," which signifies a tapster or pourer out of liquor.

The astronomical symbol of the sign ♒, representing undulating lines of waves, is said to have been the hieroglyph for water. The faint stars that seem to trail southward from the water-jar are many of them in pairs and triples, thus bearing out a stellar resemblance to a flowing stream.

In this region of the sky the 25th Hindu lunar station was situated. The Hindu name for it signified "having a hundred physicians," and it included a hundred stars, the brightest being λ Aquarii. The regent of the asterism was Varuna, the god of the waters.

The Arab lunar station or manzil known as "the felicity of tents" was also located in this region of the heavens, and the early Christians saw in this constellation the figure of St. Jude.

Aquarius, in spite of the importance attached to it by the ancients, is an inconspicuous constellation. It is characterised by a "Y"-shaped figure representing the water-jar, composed of the stars γ, ζ, η, π Aquarii. This figure was called Situla or Urna by the Latins. A rough map of South America and a rude dipper are also to be traced out in the stars of this constellation.

Alpha Aquarii is but one degree south of the celestial equator. It was called "Sadalmelik" by the Arabs, which means "the fortunate star of the king." This star marks the Chinese lunar station or Sieu, which they knew as "Goei."

The star Beta Aquarii was called by the Arabs "Sadal Sud," "the luckiest of the lucky," a title supposed to refer to the good fortune attending the passing of winter. This star and ξ Aquarii constituted the Persian lunar station known as "Bunda." On the Euphrates Beta Aquarii was known as the "star of mighty destiny."

The star Delta Aquarii marks the radiant point of the meteors known as the Delta Aquarids which appear from the 27th to the 29th of July, and in this vicinity Mayer, in 1756, noted as a fixed star the object that was later identified by Sir William Herschel as the planet Uranus.

ξ Aquarii is a double, the two suns revolving in 1624 years. They present a fine sight in a small telescope.

Piscis Australis
The Southern Fish

PISCIS AUSTRALIS
THE SOUTHERN FISH

AQUARIUS is so closely identified with the constellation Piscis Australis, or the "Southern Fish," situated directly south of it, that a description of this asterism is worthy of notice in this place.

Piscis Australis, says Burritt, is supposed to have taken its name from the transformation of Venus into the shape of a fish, when she fled terrified at the horrible advances of the monster Typhon. It has been thought that the Southern Fish was the sky symbol of the god Dagon of the Syrians, the Phagre and Oxyrinque adored in Egypt, and it has even been associated with the still greater Oannes. It was especially mentioned by Avienus as the "Greater Fish," and Longfellow in the notes to his translation of the Divine Comedy, called it the "Golden Fish."

The Mosaicists held the asterism to represent the Barrel of Meal belonging to Sareptha's widow, but Schickard pronounces it to be the Fish taken by St. Peter with a piece of money in its mouth.

Aratos describes the figure as "on his back the Fish," but it generally appears in an upright position with mouth agape, drinking in the great stream which flows down the sky from the water-jar of Aquarius.

In the early legends the Southern Fish was the parent of the Northern and Western Fishes that make up the zodiacal constellation Pisces.

This constellation as a whole is inconspicuous in this hemisphere owing to its low position. Its lucida however, the brilliant first magnitude star Fomalhaut, rises well above the horizon and adorns the southern skies in the early evening during the autumn months. Fomalhaut is made the more conspicuous because it is the brightest star in this region of the sky. It is the farthest south of all the first magnitude stars we see, and ranks thirteenth among the brilliant stars in our hemisphere.

Mrs. Martin[1] thus refers to this great sun: "On early acquaintance the loneliness of the star, added to the sombre signs of approaching autumn, sometimes gives one a touch of melancholy, but its aspect when more familiar soon comes to suggest only sweetness and serenity, and a lover of Fomalhaut feels that a sustaining light has gone when, during the last of December, this beautiful star sinks gently down in the south-west and disappears from the evening sky not to return for more than seven months."

Fomalhaut is always associated in the mind of the star lover with Capella, the brilliant in the constellation Auriga, which rises far from it over the north-eastern horizon. As these two stars rise almost simultaneously, one naturally turns from a glimpse of one to the bright beams of the other.

The name Fomalhaut, pronounced Fō′-mal-ō, is from the Arabic, meaning "the Fish's Mouth." Aratos mentions it as "One large and bright by both the Pourer's feet." Among the early Arabs, Fomalhaut was known as "the First Frog."

Flammarion tells us that Fomalhaut was known as "Hastorang" in Persia 3000 B.C., when near the winter solstice. It was also called "the magnificent Royal Star," and was one of the four Royal stars of astrology, ruling over the four cardinal points of the heavens, the other stars being Regulus, Antares, and Aldebaran. These four stars were also regarded as the four guardians of Heaven, sentinels watching over the other stars.

About 500 B.C. Fomalhaut was the object of sunrise

[1] *The Friendly Stars*, by Martha Evans Martin.

worship in the temple of Demeter at Eleusis. With astrologers it portended eminence, fortune, and power. Its position in the heavens has been determined with the greatest possible accuracy to enable navigators to find their longitude at sea, and it appears in the Ephemerides of all modern sea-going nations. It culminates at 9 P.M., on the 25th of October.

Fomalhaut is reddish in colour, and distant from the earth about twenty-one light years. So far as is known it has no companion. By one authority this star was thought to be the Central Sun of the Universe, and according to Allen no other star seems to have had so varied an orthography.

Aquila
The Eagle

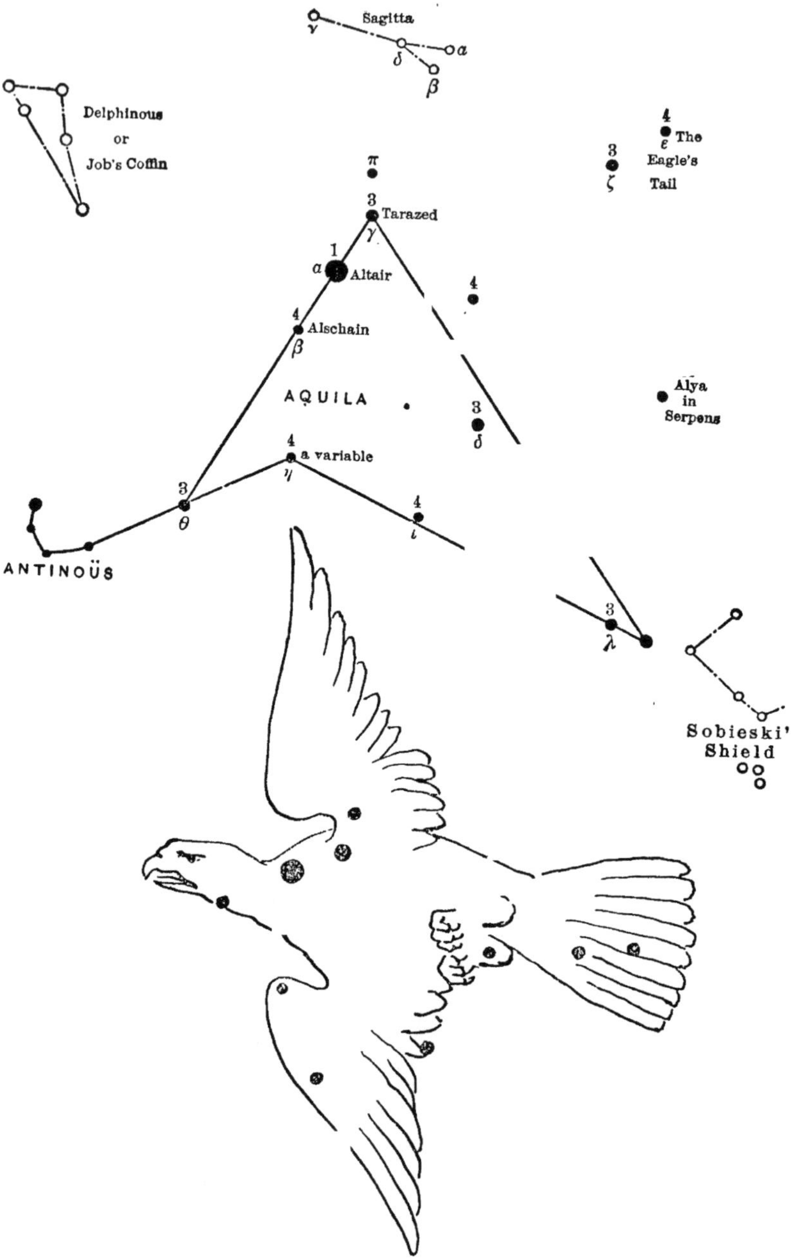
Sagitta
γ
δ
α
β
Delphinous
or
Job's Coffin
4
ε
The
3
Eagle's
ζ
Tail
π
3
Tarazed
γ
1
α
Altair
4
4
Alschain
β
AQUILA
3
δ
Alya
in
Serpens
4
a variable
η
3
θ
4
ι
ANTINOÜS
3
λ
Sobieski'
Shield

AQUILA

AQUILA
THE EAGLE

> Aquila the next
> Divides the ether with her ardent wing
> Beneath the Swan, not far from Pegasus,
> Poetic Eagle.

THE history of the constellation Aquila, the Eagle, is especially interesting both because in this case we can trace it back very clearly to the earliest times, and the original Euphratean name has been preserved.

The Sumerian-Akkadian Eagle was "Alula" (the great spirit), the symbol of the noontide sun, and in all probability the origin of the present constellation. On a Euphratean uranographic stone of about 1200 B.C., there is a bird figured, known as the Eagle, which is supposed to represent the constellation of Aquila.

The Latins knew this constellation as Aquila, and their poets called it "Jovis Ales" and "Jovis Nutrix," the "Bird" and the "Nurse of Jove." Ovid called it "Merops," King of the island of Cos, in the Archipelago, turned into the Eagle of the sky, and placed among the stars by Juno. Others thought it some Æthiopian king like Cepheus.

Aquila is generally joined with Antinoüs, an asterism invented by Tycho Brahe. Antinous was a youth of Bithynia in Asia Minor, who came to an untimely death by drowning in the river Nile. So greatly was his death lamented by the Emperor Adrian, that he erected a temple to his memory, and built in honour of him a splendid city on the banks of the Nile.

In Greece, the eagle was the bird of Zeus, and is represented as bearing aloft in his talons a beautiful boy. This youth is sometimes called Ganymede, whom Jupiter, as the story runs, desiring for his cup-bearer, sent the eagle to seize and carry up to heaven.

One of Ovid's Metamorphoses treats of Ganymede, the youthful cup-bearer, and Elizabeth Barrett Browning thus translates it in part:

> But sovran Jove's rapacious bird, the regal
> High percher on the lightning, the great eagle,
> Drove down with rushing wings; and thinking how,
> By Cupid's help, he bore from Ida's brow
> A cup-boy for his master, he inclined
> To yield, in just return, an influence kind;
> The god being honoured in his lady's woe.

Aquarius as we have seen was also supposed to represent Ganymede, and there seems to have been a connection between the constellations Aquarius and Aquila.

Another story claims that Jupiter himself assumed the form of an eagle and seized and carried off Ganymede, and Aquila was known as the bird of Jove and bearer of his thunder.

Horace thus alludes to this famous bird:

> Jove for the prince of birds decreed,
> And carrier of his thunder, too,
> The bird whom golden Ganymede
> Too well for trusty agent knew.
>
> Gladstone's translation.

Some have imagined that Aquila was the eagle which brought nectar to Jupiter, while he lay concealed in the cave at Crete to avoid the fury of his father Saturn, and this is in accordance with the legend of the Rig-Veda that Aquila bore the Soma (the invigorating juice) to India, "rushing impetuously to the vase or pitcher" (the constellation Aquarius). This legend serves to corroborate

Photo by Anderson

Ganymede Seized by the Eagle

Painting by Rubens. Gallery of the Prado, Madrid

the view that the Water Bearer and the Eagle were closely associated.

Some of the ancient poets say that this is the eagle which furnished Jupiter with weapons in his war with the giants. In accordance with this version early representations added an arrow held in the Eagle's talons. Manilius wrote:

> The tow'ring Eagle next doth boldly soar,
> As if the thunder in his claws he bore;
> He 's worthy Jove since he, a bird, supplies
> The heaven with sacred bolts, and arms the skies.

On Burritt's map, Antinoüs is represented as grasping a bow and arrows as he is borne aloft in the talons of the Eagle. In this connection there may be a significance in the position of the asterism Sagitta, the Arrow, just north of Aquila.

Among the Australians Aquila is called "Totyarguil," and represents a man who, when bathing, was killed by a fabulous animal, a kind of kelpie, as in Greece Orion was killed by a scorpion and translated to the stars.

The Hebrews know this constellation as "Neshr," an eagle, falcon, or vulture. The Arabians called it "Al-ʽOkâb," probably their black eagle. Grotius and Bayer both called the constellation "Altair," the name now borne by its brightest star.

The Turks called Aquila the "Hunting Eagle," and through all the ages it has been known as a bird of prey, the "Eagle of the Winds," the "Soaring Eagle," as contrasted with Vega near by, the "Swooping or Falling Eagle."

Here over the face of the waters as it were, just above the region of the sky known to the ancients as "the Sea," we find three birds in flight, two eagles and a swan, our Lyra, being anciently known as "the Falling Eagle." There is a significance in this arrangement that has never been satisfactorily explained. Dupuis advanced the idea that the famous three Stymphalian Birds of mythology were represented by the constellations Aquila, Cygnus, and Lyra, grouped near Hercules, whose fifth labour it was to slay them.

On the coins of many ancient countries the eagle appears. On the coinage of Sinope it is shown perched on the dolphin. In connection with the story of Ganymede, it appears on the coinage of Chalcis and Dardanos. One coin bearing the prominent stars, says Allen, was struck in Rome in 94 B.C., by Manius Nepos, and a coin of Agrigentum bears Aquila, with Cancer on the reverse,—the one setting as the other rises.

The Chinese have here the Draught Oxen mentioned in the book of Odes, compiled about 500 B.C., and strangely enough Alpha Aquilæ, or Altair, is known among the Japanese as the boy with the ox.[1]

This constellation and Lyra are associated with the curious Chinese legend of the Spinning Damsel and the Magpie Bridge, a legend current in Korea also. It is as follows: A cowherd fell in love with the spinning damsel. Her father in anger banished them both to the sky, where the cowherd became α, β, and γ Aquilæ, and the spinning damsel the constellation Lyra. The father decreed that they should meet once a year, if they could contrive to cross the river (the Milky Way). This they were enabled to do by their friends the magpies, who still once a year, the seventh night of the seventh moon, congregate at the crossing point, and form a bridge for them to pass over. In Korea if a magpie is seen about its usual haunts at this time the children stone it for shirking its duty. According to Lafcadio Hearn, this legend is the basis of the Japanese festival called "Tanabata." The sky lovers here are known as "the Herdsman and the Weaver," and when the meeting occurs it is said that the lover stars burn with five different colours. If rain falls at the time set

[1] The early Christians likened this figure to St. Katharine and the Standard of Rome.

Photo by Hollyer

Ganymede

Painting by George Frederick Watts

for the crossing, the meeting fails to occur. For this reason rain on the Tanabata night is called the rain of tears.

Dr. Seiss regards Aquila as symbolical of the Wounded Prince or Christ suffering for mankind.

Aquila contains a star of the first magnitude called "Altair," α Aquilæ, to which Mrs. Martin in her delightful book, *The Friendly Stars*, thus charmingly refers: "Then there comes a soft June evening with its lovely twilight that begins with the last song of the woodthrush and ends with the first strenuous admonitions of the whip-poorwill, and almost as if it were an impulse of nature one walks to the eastern end of the porch and looks for Altair. It is sure to be there, smiling at one just over the tree-tops with a bright companion on either side, the three gently advancing in a straight line as if they were walking the Milky Way hand in hand and three abreast."

Allen tells us that the name of this beautiful star is from a part of the Arabic name for the constellation, and means the flying vulture.

Ovid thus alludes to the rising of Altair:

> Now view the skies
> And you 'll behold Jove's hook'd-bill bird arise.

This star was ill omened in astrology, and supposed to portend danger from reptiles. It is an important star for the mariner, however, as the moon's distance is taken from it for computing longitude at sea.

According to Dr. Elkin, Altair is fifteen light years distant from the earth. It is said to be approaching the earth at the rate of twenty-seven miles per second, and culminates at 9 P.M. on the 1st of September.

The radiant point of the meteors known as the Aquilids, visible from June 7th to August 12th, is located about five degrees east of Altair. Strangely enough in the year 389 A.D., a famous temporary star, or comet, appeared in this vicinity. Cuspinianus stated that it equalled

Venus in brilliancy. It vanished after three weeks' visibility.

Altair with its two companions Beta and Gamma Aquilæ constitute the so-called "Family of Aquila." The line joining these stars is five degrees in length. In China these stars were known as "Ho Koo," meaning "a river drum," and the Persians-called them the "Star Striking Falcon." They formed the 23d Hindu lunar station known as "the Ear." The regent of the asterism is Vishnu, and these three stars represent the three steps with which Vishnu is said in the early Hindu mythology to have strode through heaven. A trident is often given as the figure of this group.

Eta Aquilæ is a remarkable variable star. Its greatest brightness continues but forty hours. It then gradually diminishes for sixty-six hours, when its lustre remains stationary for thirty hours. It then waxes brighter and brighter until it appears again as a star of the third magnitude. From these phenomena, says Burritt, it is inferred that it not only has spots on its surface like our sun, but that it also turns on its axis. The spectrum of this star is similar to that of our sun. Lockyer thinks it is a spectroscopic binary, that is a star with a companion too close to be revealed by the most powerful telescope.

Aries
The Ram

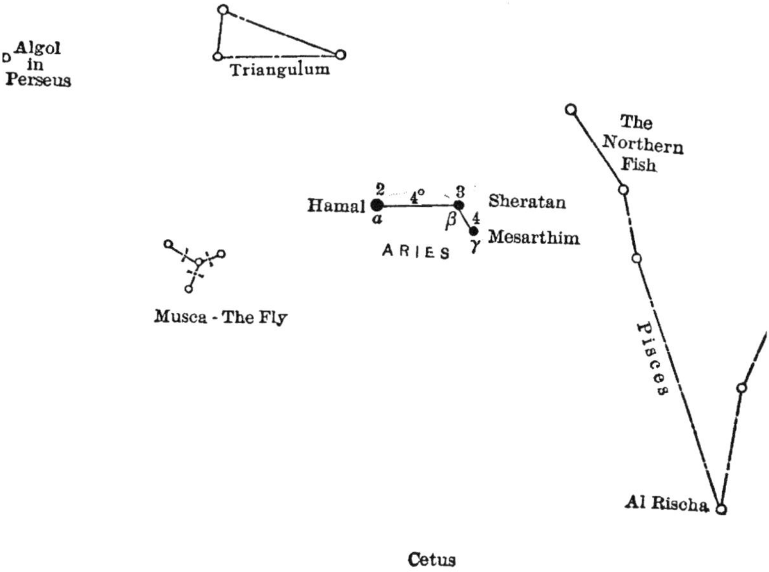
Algol
in
Perseus
Triangulum
The
Northern
Fish
Hamal
Sheratan
Mesarthim
2
4°
3
4
a
β
γ
ARIES
Musca - The Fly
Pisces
Al Rischa
Cetus

ARIES

ARIES
THE RAM

First from the east, the Ram conducts the year;
Whom Ptolemy with twice nine stars adorns,
Of which two only claim the second rank,
The rest, when Cynthia fills the sign, are lost.

Aries has been called the "Prince of the Zodiac," the "Prince of the Celestial Signs," and the "Leader of the Host of the Zodiac." It has also been associated with the ram into which Zeus changed himself to escape the pursuit of the giants. He fled to Egypt, and there the constellation was called "Jupiter Ammon."

In Chaldea, where the constellation is supposed to have originated, the ram simply represents the favourite animal of the shepherds. Considering the fact that Aries is in an inconspicuous part of the heavens, and comprises only three stars of any importance, it is surprising the wealth of lore and legend that surrounds it, and the attention paid to it by the ancients, unless we attribute to it some extraneous claim for notoriety, such as the position of these stars as regards the sun at a certain period of the year. There is little doubt that this is the real cause of the importance of this constellation.

"If," says Plunket, "we find Aries equally honoured by several nations in very early times, either these nations, independent of each other, happened to observe and mark out the sun's annual course through the heavens at exactly the same date, and therefore chose the same date, or we must suppose that they derived their calendar and knowledge of the zodiac from observations originally made by some one civilised race."

It is easy to see, as Brown avers, that the comparison of the sun to a ram or bull is a line of thought which naturally and spontaneously arises in the mind of archaic man.

In the Euphratean Valley, the probable birthplace of the constellations, the sun was styled a "Lubat," meaning old sheep, and ultimately the planets were called "old sheep stars." Hence the symbolic view of the sun as an old sheep or ram is necessarily of a remote antiquity.

In Aries we have very clear proof that many of the constellations must be regarded as mere symbols, and in nowise to be thought of as owing their names to a fancied resemblance to some creature or object, for the obtuse angle formed by the three principal stars in Aries could only resemble at best the hind leg of a sheep or ram, and so we are bound to the conviction that the ram is simply a symbol.

One theory holds that the solar ram, the sun who opened the day, was in time duplicated by the stellar ram, who in 2540 B.C. opened the year, and "led the starry flock through it as their bell-wether."

Unfortunately for this theory, as Maunder points out, we know that the constellations were mapped out at a far earlier epoch, when the equinox fell not in Aries, but in the middle of the constellation Taurus.

In mythology Aries has always represented the fabled ram with fleece of gold. Manilius thus describes it:

> First Aries, glorious in his golden wool,
> Looks back and wonders at the mighty Bull.

The old fable is as follows: Phrixus and Helle were children of Athamas, the legendary King of Thessaly. Their step-mother treated them with such cruelty that Mercury took pity on them, and to enable them to escape their mother's wrath sent a ram to bear them away. Mounted on the ram's back the children sped over land

and sea, but unfortunately Helle neglected to secure her hold, and fell from her seat while the ram was flying across the strait which divides Europe from Asia. In memory of this catastrophe this strait was afterwards known as the Hellespont. Manilius thus refers to this episode:

> First golden Aries shines, who whilst he swam
> Lost part of 's freight and gave to sea a name.

Longfellow also alludes to Helle's fall:

> The Ram that bore unsafely the burden of Helle.

Phrixus landed safely at Colchis, at the eastern end of the Black Sea. Out of gratitude for his safe deliverance, he sacrificed the ram and gave the golden fleece to the king of the country, who hung it in the sacred grove of Ares, under guard of a sleepless dragon.

The golden fleece has always been associated in Greek mythology with the voyage of the ship *Argo*, and the celebrated Argonautic expedition which set forth in search of it.

The theory has been advanced that the stellar symbols were intended simply as a record of this famous expedition. Even so good an astronomer as Sir Isaac Newton held this view, but Maunder on the contrary claims that there was nothing in the story of the neighbouring constellations to support the legend of the golden fleece.

Curiously enough Aries is the leading sign in all the systems of astrology which have come down to us through the Greeks, and it figures as the leading sign in most of the explanations of the constellation figures which are on record. Maunder considers that this fact proves that these astrological systems, and these theories concerning the constellation figures, not only took their rise at a later epoch, but that when they did so, the real origin and meaning of the designs had been wholly lost.

One peculiar fact respecting Aries for which there is no apparent explanation, is that the ram is always represented with reverted head. On a coin type of Cyzicus, about 500–450 B.C., the ram is thus depicted. Allen notes as an exception to this almost universal figure, the ram erect in the *Albumasar* of 1489.

Berosus, a Babylonian priest in the time of Alexander the Great, said that the ancients—those ancient to him—believed that the world was created when the sun was in Aries.

Pliny said that Cleostratos of Tenedos first formed Aries, but there is no doubt that the constellation originated many centuries before this.

Plunket informs us that in the Egyptian calendars no reference is made to Aries, but in Egyptian mythology the importance of the ram is revealed. Amen or Amon, the great god of the Theban triad, is sometimes represented as ram-headed. The great temple to him in conjunction with the sun, *i.e.*, to Amen-Ra, is approached through an avenue of gigantic ram-headed sphinxes. At the season of all the year when Aries specially dominated the ecliptic, the statue of the god Amen was carried in procession to the Nekropolis, from which place the constellation Aries was fully visible. "The preparations for this great festival began before the full moon next to the spring equinox, and on the fourteenth day of that moon all Egypt was in joy over the dominion of the Ram. The people crowned the Ram with flowers, carried him with extraordinary pomp in grand procession, and rejoiced in him to the utmost." The ancient Persians, who called Aries "Bara," had a similar festival.

Between 1400 and 1100 B.C., when Rameses II. dedicated the temple of Aboo Simbel, the sun when it penetrated into the shrine of the temple was in conjunction with the first stars of the constellation Aries, and this fact doubtless led the King to honour Aries in connection with the god Amen. The Egyptians called Aries "the Lord of the Head."

Avenue of Ram-headed Sphinxes, Karnak
From Piers's " *Inscriptions of the Nile Monuments*"

Not only the Egyptians, but all the great civilised nations of the East, had traditions of a year beginning when the sun and moon entered the constellation Aries.

Jensen is of the opinion that Aries may have been first adopted into the zodiac by the Babylonians when its stars began to mark the vernal equinox. Plunket on the contrary, thinks that the choice of the constellation as Prince and Leader of the signs was made, not when its stars marked the spring equinox, but when they indicated the winter solstice. According to this view Aries, Cancer, Libra, and Capricornus marked the four seasons and the cardinal points in 6000 B.C.

In the Rig-Veda, the first lunar station in the Indian series is named "Aswini." The two chief stars in the station are the twin stars as they may be called, β and γ Arietis. Joyous hymns were addressed to the twin heroes, the Aswins, which may properly be called New Year's hymns, composed in honour of these stars, whose appearance before sunrise heralded the approach of the great festival day of the Hindu New Year. Next to Agni and Soma, the twin deities named the Aswins are the most prominent in the Rig-Veda. They are celebrated in more than fifty entire hymns, while their name occurs more than four hundred times. These twin heroes of Hindu mythology correspond to the famous twins of Grecian mythology, Castor and Pollux.

The Arabs, whose first manzil or lunar station was formed by these same two stars, knew them as "the two tokens," that is to say of the opening year. They called the constellation Aries "Al-Hamal," the Sheep, while the early Hindus called it "Aja," and "Mesha."

The Hebrews called the constellation "Telī," and assigned it in their zodiac to either Simeon or Gad. Dr. Seiss, following Cæsius, regarded Aries as symbolising the Lamb of the World.

Aries, the April sign according to Hagar, was known in Peru as "the Market Moon" or "Kneeling Terrace."

At this season the early crops were harvested and borne home on the backs of llamas. The festival was called "Ayri huay" or that of the axe, and referred to the reaping of these crops. This conception of the constellation is decidedly at variance with the Eastern idea of it.

The Syrians called Aries "Amru" or "Emru," while the Turkish name for the constellation was "Kuzi."

The Romans generally called the constellation "Aries," but Ovid named it "Phrixea Ovis" and "Cornus." Other Latin names for it are "Vernus Portitor," the Spring-bringer, and "Arcanus."

As one of the zodiacal twelve of China, Aries was first known as "the Dog," and later as "the White Sheep." At the time when it was sought to reconstruct the constellations on Biblical lines, Aries was selected to represent Abraham's ram caught in the thicket, or St. Peter.

The Anglo-Normans of the 12th century called Aries "Multuns," and the poet Dante refers to it as "Montone."

In Italy, France, and Germany, Aries is called respectively, "Ariete," "Bélier," and "Widder." The symbol of the constellation ♈ probably represents the head and horns of the animal. In this region of the sky a brilliant temporary star appeared in the year 1012 A.D.

Astrologically considered Aries is the house and joy of Mars, and signifies a dry constitution, long face and neck, thick shoulders, swarthy complexion, and a hasty passionate temper. It governs the head and face, and all diseases relating thereto. It reigns over France, England, Germany, Switzerland, Denmark, Lesser Poland, Syria, Naples, Capua, Verona, etc. It is a masculine sign, and is regarded as fortunate.

According to Eleanor Kirk, who is a great authority on the subject, people born under Aries, that is between Mar. 20th and Apr. 19th are usually very executive, earnest, and determined. They are leaders, and dominate those about them. They are noble, generous, progressive, and have occult power. They are good scholars, bright, genial, and witty.

The natal gem of Aries is the bloodstone, the symbol of good luck; the natal flower, the violet; the metal, iron.

Alpha Arietis was called "Hamal" or "Hamel" by the Arabs, meaning a sheep, and the name "Al-Nath" has also been found for it on some of the ancient Arabic globes. "Arietis" is another name for this star.

Among the Greeks in early times, Hamal held the important office of sunrise herald, at the vernal equinox. In Ptolemy's list it is described as "The one above the head" (of the Ram), and astrologers regarded it as dangerous and evil, denoting bodily hurts.

Brown asserts that the stellar Ram was in the first place only the star Hamal, the constellation being formed around it afterwards. Chaucer refers to the star as "Alnath," that is to say the "horn push," a name more commonly associated with the star in the tip of the northern horn of the Bull, a star common to the constellations Taurus and Auriga.

Other Euphratean names for this star have been "Lu-lim" or "Lu-nit," the ram's eye, and "Simal," the Horn Star. It was also called "Anuv" and "Ku," meaning the Prince or the Leading One, the ram that led the heavenly flock.

Of the Grecian temples, at least eight, at various places, and of dates ranging from 1580 to 360 B.C. were oriented to this star, and it is the only star to which Milton makes individual allusion.

Hamal is much used in navigation in connection with lunar observations, and culminates at 9 P.M. on the 11th of December. It is approaching our system at the rate of nine miles per second. According to Miss Clerke,[1] Hamal is distant from the earth about forty light years.

The star Beta Arietis was known to the Arabs as "Shara-

[1] *The System of the Stars*, by Agnes M. Clerke.

tan," meaning "a sign," this star having marked the vernal equinox in the days of Hipparchus.

Gamma Arietis has been called the "First star in Aries" as at one time it was nearest to the equinoctial point. It is a beautiful double star, easily visible in a small telescope, and was discovered to be double by Dr. Hooke in 1664. This star was known to the Arabs as "Mesarthim," meaning the "two attendants," a reference to Beta and Gamma Arietis, these two stars being considered as attendants on Hamal. The Persians called these stars "The Protecting Pair."

The faint stars east of Hamal on the back of the Ram form a little group known as "Musca Borealis," the Northern Fly. The figure appears in Burritt's Atlas. According to Allen the inventor of this asterism is unknown. Musca has been also styled "the Wasp" and "the Bee." It comes to the meridian on the 17th of December at 9 P.M.

Auriga
The Charioteer

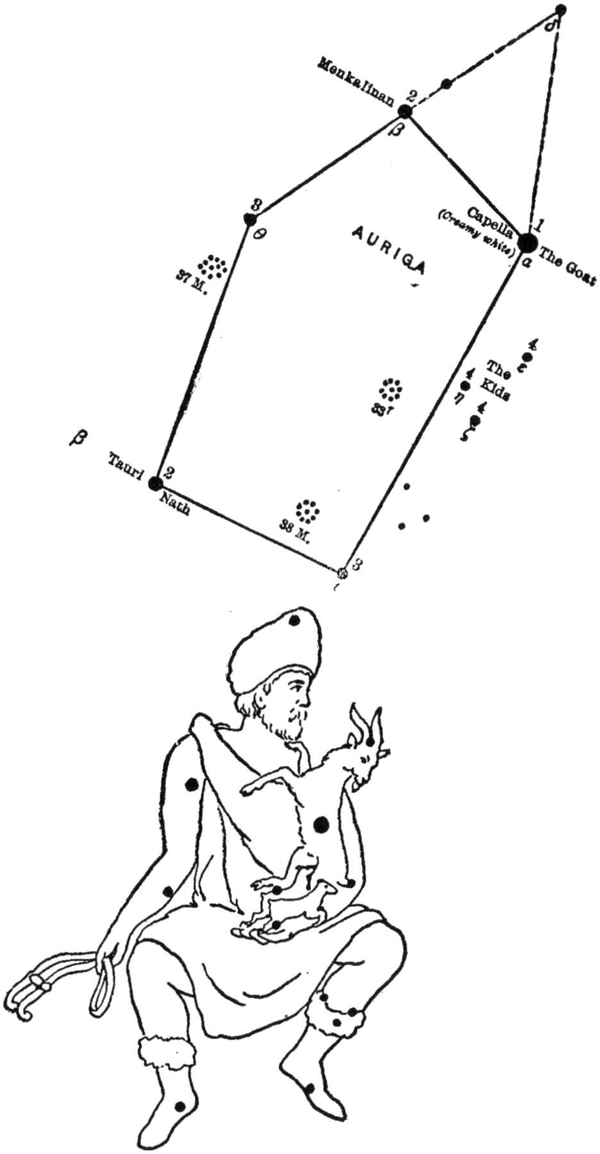

Menkalinan
Capella
(Creamy white)
The Goat
AURIGA
37 M.
The Kids
Tauri
Nath
38 M.

AURIGA

AURIGA
THE CHARIOTEER

Close by the kneeling Bull behold
The Charioteer who gained by skill of old
His name and heaven as first his steeds he drove,
With flying wheels, seen and installed by Jove.

MANILIUS.

THE origin of this ancient constellation is lost. It has been represented for ages as a mighty man seated on the Milky Way, and like a shepherd carrying a goat on his shoulder, and a pair of little kids in his hand. The first magnitude star Capella shines in the heart of the imaginary goat.

Allen says: "The results of modern research give us reason to think that this constellation originated on the Euphrates, in much the same form as we have it to-day. It certainly was a well established sky figure there millenniums ago. A sculpture from Nimroud is an almost exact representation of Auriga, with the goat carried on the left arm."

On the Assyrian tablets Auriga was the "Chariot," and in accordance with this in Græco-Babylonian times the constellation "Rukubi," the Chariot, lay here nearly coincident with our Charioteer.

Seen rising in the north-east, it needs but little imagination to trace in the stars of Auriga a resemblance to an ancient Roman chariot, so that the title "Chariot" seems more appropriate than "Charioteer."

Ideler thinks that the original figure was made up of the five stars α, β, ε, ζ, η. The driver (represented by the star Capella) is imagined as standing on an antique sloping chariot, marked by β. The other stars represent the reins. The illustration, although contrary to Ideler's conception, seems a much easier figure to trace. Here as in Ideler's figure Capella represents the driver's head. (See p. 69.)

Plunket suggests 3000 B.C. as the date of the invention of the constellation Auriga, for then Capella, the brightest star in this region of the sky, was on the meridian in conjunction with the sun at noon of the spring equinox, and in opposition at midnight of the autumnal equinox.

Capella has by several writers been identified with the star "Icu of Babylon," mentioned in many of the Babylonian texts, and the star of Marduk. If this is correct we should credit the Babylonian astronomers with the delineation of the figure of Auriga.

Auriga has also been identified with Erichthonius, the son of Vulcan and Minerva, who being deformed and unable to walk invented the chariot, an achievement that secured him a place in the sky.

> Bold Erichthonius was the first who join'd
> Four horses for the rapid race designed,
> And o'er the dusty wheels presiding sate.
>
> Dryden.

Swinburne sings of this famous inventor in the following lines:

> Thou hast loosened the necks of thine horses, and goaded their flanks with affright,
> To the race of a course that we know not on ways that are hid from our sight;
> As a wind through the darkness the wheels of their chariot are whirled,
> And the light of its passage is night on the face of the world.

Manilius thus refers to the Charioteer

> Near the bent Bull a seat the Driver claims,
> Whose skill conferr'd his honour and his names.

His art great Jove admired, when first he drove
His rattling Car, and fix't the Youth above.

According to Lemprière, Erichthonius became the constellation Boötes instead of Auriga.

Brown identifies Erichthonius with Poseidon, the lord of the abyss below the surface of the earth, the stormy earth-shaking divinity, and thus accounts for the stormy influence that the Greeks attributed to Capella, the Goat-star.

The Greek name for Auriga was Ἑνιοχος, "the holder of the reins," a name preserved for us in the Arab name for Beta Aurigæ, "Menkalinan," meaning "the shoulder of the rein-holder."

Blake[1] thinks that the proximity of the chariot (Ursa Major) accounts for the name of the Charioteer applied to the constellation.

On a French chart of 1650 Auriga figures as Adam with his knees on the Milky Way, and the she-goat climbing over his neck.

Dr. Seiss claims that Auriga represented to the Greeks the Good Shepherd, a symbol foretelling the coming of Christ.

Cæsius likened it to Jacob deceiving his father with the flesh of his kids.

Auriga has also been identified with Myrtilus, the charioteer of Œnomaus, with Cillas, Pelethronius, Hippolytus, Bellerophon, and St. Jerome, while Jamieson is of the opinion that Auriga is a mere type or scientific symbol of the beautiful fable of Phaëton, because he was the attendant of Phœbus at the remote period when Taurus opened the year.

Auriga in its glorious lucida Capella contains a star famous in the history of all ages. To the early Arabs Capella was known as the "Driver," because it appears in the evening twilight earlier than the other stars, and so apparently

[1] *Astronomical Myths* by J. F. Blake.

watches over them, or still more practically as the "Singer" who rides before the procession cheering on the camels, which last were represented by the Pleiades. They also called it "the Guardian of the Pleiades."

Capella is a singularly beautiful object, and lies nearer the Pole than any other of the first magnitude stars. It rises almost exactly in the north-east, and July is the only month in the year when it is not visible in these latitudes sometime before midnight.

Seen in the cool evenings of early fall, flashing its wonderful prismatic rays from the low eastern sky, it seems like a herald of old announcing the coming of a mighty host, the brilliant stellar pageant that graces our clear winter nights, and renders them gorgeous with light and life.

Mrs. Martin thus refers to the rising of this famous star, a star which Tennyson designates as "a glorious crown": "When you watch the birds congregating in noisy flocks in the morning for the fall migration, and in the afternoon gather the first fringed gentians, look for Capella in the north-eastern sky in the evening . . . the fair, golden, bright Capella, that decks the sky in its season. We follow it in its course visible to us across the heavens, we joy in its beauty, and feel the kindly influence that astrologers have always ascribed to it."

Eudosia thus alludes to the brilliance of Capella:

> And scarce a star with equal radiance beams
> Upon the earth.

Capella means "the little she-goat," the goat which suckled the infant Jupiter. The story runs that having in his play broken off one of the goat's horns, Jupiter endowed the horn with the power of being filled with whatever the possessor might wish, whence it was called "the Cornucopia," or "horn of plenty." This title is also applied to the horn of Capricornus the Sea Goat.

In India, Capella was worshipped as the Heart of Brahma. The ancient Peruvians called it "Colca," and connected it with the affairs of shepherds. English poets have alluded to it as "the Shepherds' Star." These allusions have reference doubtless to the time of Capella's culmination, which corresponded with the season when the shepherds watched their flocks.

Probably the oldest allusion to Capella extant is that which was found on an old tablet in Akkadian, which has been translated as follows: "When on the first day of the month Nisan the star of stars (or Dilgan) and the moon are parallel, that year is normal. When on the third day of the month Nisan the star of stars and the moon are parallel, that year is full."

"The star of stars" of the inscription, says Maunder, is no doubt Capella, and the year thus determined by the setting together of the moon and Capella would begin on the average with the spring equinox about 2000 B.C. The date of the Akkadians is about 4000 years ago.

Allen tells us that Capella's place on the Denderah zodiac is occupied by a mummied cat in the outstretched hand of a male figure crowned with feathers. While always an important star in the temple worship of the great Egyptian god Ptah, the Opener, it is supposed to have borne the name of that divinity, and probably was observed at its setting 1700 B.C. from his temple, the noted edifice at Karnak near Thebes. Another recently discovered sanctuary of Ptah, at Memphis, was also oriented to Capella. Lockyer thinks at least five temples were oriented to its setting.

A stormy character has been attributed to Capella, and hence it has sometimes been called "the rainy Goatstarre." Aratos alludes thus to its stormy influences:

> Capella's course admiring landsmen trace,
> But sailors hate her inauspicious face.

Similarly the poet Callimachus who lived about 240 B.C. wrote:

> Tempt not the winds forewarned of dangers nigh,
> When the kids glitter in the western sky.

The Kids are represented by the three fourth magnitude stars, ε, ζ, and η Aurigæ, which form a small isosceles triangle close to Capella and serve to identify that star. They were sometimes called "the stormy Hædi," and were so much dreaded as presaging the stormy season on the Mediterranean, that their rising early in October evenings was the signal for the closing of navigation.

All classical authors who mentioned the stars, says Allen, alluded to the direful influence of Capella, and a festival, the "Natalis Navigationis," was held when the days of that influence were past.

Astrologically Capella portended civic and military honours, and wealth.

Some astronomical facts relative to Capella may be of interest. Capella in its spectrum almost exactly resembles the sun. It is a spectroscopic binary, its duplicity being alone revealed by the spectroscope. Its period of revolution is 104 days, and its unseen companion has a spectrum resembling that of Procyon, a star further advanced in the order of development than Capella.

In brightness Capella ranks third of all the stars we see in these latitudes, and fifth of all the stars in the firmament. Its mass is eighteen times that of the sun.

Dr. Elkins gives its parallax, that is its distance from the earth, as approximately thirty-four light years. A light year is the distance light travels in one year, at the terrific speed of 186,000 miles a second.

Capella is receding from the earth at the rate of about fifteen miles a second, and in about 3,000,000 years will appear as a second magnitude star.

The Temple of Khonsu, Karnak
From Piers's "*Inscriptions of the Nile Monuments*"

Ptolemy, El Fergani (10th century), and Riccioli have all called Capella red.

If the earth were midway between Capella and the sun, we should receive 250 times as much light from Capella as from our little solar star. According to Newcomb, Capella is 120 times as bright as the sun, and the sun at the distance of Capella would appear as a 5.5 magnitude star. The star culminates at 9 P.M. on Jan. 19th.

"Beta Aurigæ is supposed to be a very close binary. The two practically equal stars that compose the pair are estimated to be only seven and one half millions of miles apart, and revolving in a period of about four days with a relative velocity of fully 150 miles a second according to Prof. Pickering. It is receding from the earth at the rate of about seventeen miles a second."

Gamma Aurigæ, called by the Arabs "Al-Nath," is common to the constellations Taurus and Auriga, and marks the tip of the Bull's right horn.

The remaining stars in the constellation call for no special comment, but Auriga is rich in star clusters, M. 37 being especially noteworthy. Smith calls this "a magnificent object, the whole field being strewed, as it were, with sparkling gold dust; and the group is resolvable into about 500 stars. Even in small instruments this cluster is extremely beautiful, one of the finest of its class."

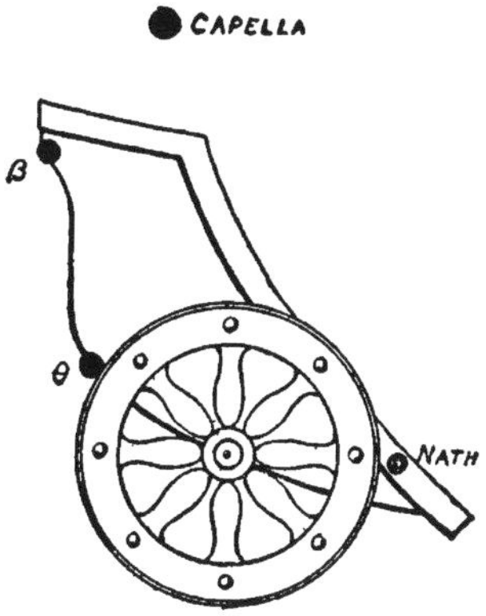

Boötes
The Bear Driver

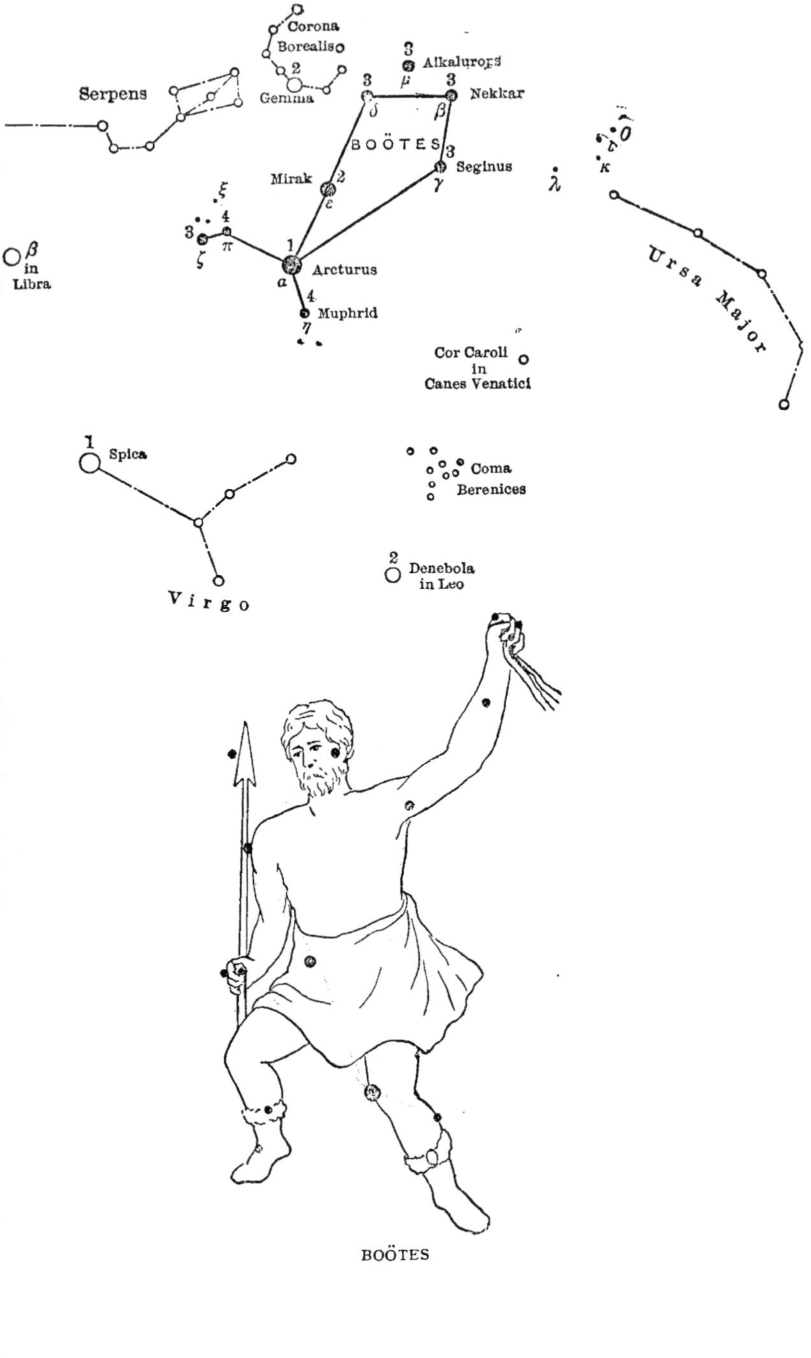
Corona
Borealis
2
Gemma
Serpens
3
Alkalurops
μ
3
Nekkar
3
δ
β
BOÖTES
3
Seginus
γ
Mirak
2
ε
λ
κ
ι
θ
ξ
4
3
π
ζ
β
in
Libra
1
Arcturus
α
4
Muphrid
η
Ursa Major
Cor Caroli
in
Canes Venatici
1
Spica
Coma
Berenices
2
Denebola
in Leo
Virgo

BOÖTES

BOÖTES
THE BEAR DRIVER

And next Boōtes comes whose ordered beams
Present a figure driving on his teams.

MANILIUS.

THE original title of this constellation was in all probability "Arcturus," the present title of the lucida of the constellation, a famous star of the first magnitude. The title Boōtes, pronounced Bō-ō′-tēz, appeared in the *Odyssey*, and according to Allen has been in use for at least three thousand years.

The stars in this region of the sky seem to have attracted the admiration of almost all the eminent writers of antiquity. Aratos pays this tribute to Boōtes:

Behind and seeming to urge on the Bear
Arctophylax, on earth Boōtes, named
Sheds o'er the arctic car his silver light.

And eight hundred years later Claudian wrote:

Boōtes with his wain the north unfolds.

Boōtes is represented by the figure of a mighty man with uplifted hand, holding in leash two hunting dogs. He seems to be pursuing the Great Bear around the Pole, and hence Boōtes is often referred to as "the Bear Driver."

Carlyle in *Sartor Resartus* thus mentions the constellation: "What thinks Boōtes of them as he leads his Hunting Dogs over the zenith in their leash of sidereal fire?"

Boōtes is also represented as a Herdsman and a Ploughman, guiding the Wain, as the constellation Ursa Major is sometimes called. Cicero takes this view, and adds that Boötes was sometimes called "Arctophylax" from two Greek words signifying "bear keeper" or "bear driver."

The name Boötes, according to some authorities, is derived from the Greek βους, meaning ox, and ὠθεῖν, to drive. Others claim that the title was derived from βοητής, meaning clamorous, descriptive of the shouts of the driver, or the call of encouragement to the hounds, hence the constellation has been sometimes called "Vociferator" and "Clamator."

The mythology of the constellation is interesting. According to some of the Greeks it represented Icarius the father of Origone, others claim it represented Erichthonius, the inventor of the chariot. It was also said to be Arcas, the son of Zeus and the nymph Callisto.

Plunket claims the date 6000 B.C. and latitude 45 degrees north, for the time and place of the invention of this constellation, as then and there Boötes might be seen at midnight of the summer solstice standing upright on the northern horizon, his head reaching nearly to the Pole. Never since that date has he held so commanding a position in the sky, nor at any more southern latitude could his whole figure have been represented as standing on the horizon.

Boötes has also been called "Atlas" from its nearness to the Pole, and because it appeared to hold up the heavens. In all probability Boötes has been deprived of an arm, the stars formerly representing it now forming the constellation of the Northern Crown. Proctor thinks that this change was made at some time preceding that of Eudoxus, who was born about 300 B.C.

The risings and settings of Boötes which took place near the equinoxes portended great tempests. Boötes sets in a perpendicular position, and takes eight hours to make his exit, hence allusions to his sluggish and tardy movements

Photo by Brogi

Atlas
National Museum, Naples

are found in the works of the ancients. Manilius thus refers to this peculiarity of the Herdsman:

> Slow Boötes drives his ling'ring teams.

And Aratos describes him as:

> When tired of day
> At even lingers more than half the night.

Boötes is an early riser so to speak, making up for his late hours, as he rises horizontally, "all at once," as Aratos wrote.

According to Allen the early Catholics knew Boötes as St. Sylvester. Cæsius said it might represent the prophet Amos, the Herdsman or Shepherd, and Dr. Seiss thought it represented the Great Shepherd and Harvester of Souls.

The shepherd idea as connected with Boötes is borne out by its proximity to the Pole, which the Arabs regarded as a sheepfold, and Boötes has accordingly been called "Pastor" by some, meaning Shepherd. This title conforms to the title "Sibzianna" for the constellation, which appears on the ancient Euphratean star list, and which means "Shepherd of the Life of Heaven."

Burritt informs us that the ancient Greeks called this constellation "Lycaon," a name derived from λυχος which signifies "a wolf." The Hebrews called Boötes "Caleb Anubach," meaning "the Barking Dog," while the Latins among other names called it "Canis."

This allusion to a barking dog and a wolf in connection with Boötes seems to refer again to the Arabs' polar sheepfold. Their imaginary picture contained a flock of sheep, a shepherd and his dog, and a wolf or hyena lurking near by in search of prey.

"Seginus," "Nekkar," and "Alkalurops" are names that have also been applied to this constellation, but now they appear as individual star names.

Boötes also figures as a spear or lance bearer, the shepherd's staff which he was represented as bearing having been changed into a more formidable weapon.

In Burritt's Atlas Boötes appears with his back turned to the bear which his hounds are closely following, and his attitude is anything but one of pursuit.

Landseer and Lalande both held that the Bear Driver was the national sign of ancient Egypt, the myth of the dismemberment of Osiris originating in the successive settings of its stars, and that there it was called "Osiris," "Bacchus," or "Sabazius," the ancient name for Bacchus and Noah.

The star Alpha Boötis bears the name "Arcturus." This glorious star has excited the admiration of all mankind, and from the earliest times we find it mentioned. Without doubt it was one of the first stars to be named. Arcturus is one of the few stars alluded to in the Bible, where we find a reference to it in the Book of Job, hence it is sometimes called "Job's star."

Arcturus probably owes its name to its proximity to Ursa Major, as it means "the watcher of the Bear." The name of this star according to Gore is derived from the Greek words ἄρκτος and οὐρά, which signify a bear's tail, so called because it lies nearly in the continuation of the Great Bear's tail.

Virgil frequently mentions Arcturus, and Manilius in his reference to Boötes thus speaks of its position in relation to the figure of the Herdsman:

> Below his girdle, near his knees, he bears
> The bright Arcturus, fairest of the stars.

In early days Arcturus represented a spear in the hunter's hand, and with the Arabs it was "the Lance Bearer." Emerson, in his translation of the Persian poet Hafiz, wrote:

> Poises Arcturus aloft morning and evening his spear.

Like many other prominent stars Arcturus shared its name with the constellation. Miss Clerke is of the opinion that Arcturus received its name long before the constellation was thought of, forming the nucleus of a subsequently formed group.

Allen states that this star was famous with the seamen of early days even from the traditional period of the Arcadian Evander, and regulated the annual festival by its movements in relation to the sun.

Mrs. Martin thus paints a scene of springtime to which Arcturus lends its lustre: "What more gracious day's progress in beauty could there be than to travel with the eyes from the cheerful hepaticas dotting the soft ground among the trees to the round, white, silent blossoms of the dogwood fringing the late April woods, and thence, when the evening falls, to the bright yet gentle light of Arcturus in the sky, announcing the end of the purple twilight."

The Chinese designated Arcturus "the palace of the Emperors." They also called it "Ta Kiō," meaning the "Great Horn," four small stars near by being "Kang Che," the "Drought Lake."

The Eskimos called Arcturus "Sibwudli," and it is the timepiece of the seal netters during the great night fishing in December and January. The position of this brilliant star as it circles round the Pole enables them to judge how the night is passing.

The Arab name for Arcturus was "Al-simāk-al-Rāmih," meaning "the simak armed with a lance," also translated "the leg of the Lance-Bearer," and "the lofty Lance-Bearer." Gòre states that according to the Persian astronomer Al-Sufi, who wrote a description of the heavens in the 10th century, the word simak means "elevated," referring to the high altitude the star attains above the horizon. Schjellerup however, thinks that the word refers to the brilliancy of the star and not to its altitude. The Arabs also knew Arcturus as "the Keeper of Heaven."

In India, Arcturus marked the 13th lunar station, known

as "the Good Goer" or perhaps "sword," but figured as a coral bead, gem, or pearl. It was also known in India as "the outcast." As might be expected of so conspicuous a star, we find many of the Egyptian temples oriented to it.

Al-Biruni mentioned Arcturus as "the Second Calf of the Lion," the star Spica representing the First Calf. Allen states that this star has been identified with the Chaldeans' "Papsukal," the "Guardian Messenger," while according to Smith and Sayce, Arcturus was "the Shepherd of the Heavenly Flock," or "the Shepherd of the Life of Heaven," undoubtedly the Sib-zianna of the inscriptions. Strange to say the Eskimo title for the star, Sibwudli, has the same first syllable as the title of the Euphratean hieroglyphics.

Arcturus was long supposed by the ancients to be the nearest star to the earth. Its influence was always dreaded, as the writings of Aratos and Pliny testify, and its rising and setting were supposed to portend great tempests.

Hippocrates, who lived about 460 B.C., made much, says Allen of the influence of Arcturus on the human body, in one instance claiming that a dry season after its rising agrees best with those who are naturally phlegmatic, and that diseases are especially apt to prove critical in these days.

Astrologically those born under Arcturus were destined to have honour and riches conferred on them.

Arcturus is a remarkable star by reason of its rapid motion through space, indeed it may rightly be called "a runaway star." Since the days of Ptolemy it has moved over a distance equal to fully twice the moon's apparent diameter, and even to the naked eye it no longer fits the alignment with other stars which Ptolemy described. Its proper motion in miles per second is given by different authorities as anywhere from one hundred to three hundred miles.

There is also great discrepancy in the estimate of the brightness of Arcturus as compared with the sun. Prof.

Russell claims that Arcturus exceeds our sun in brilliance one hundred and fifty times, while some make Arcturus equal in illuminating power to six thousand such suns as ours.

According to Mrs. Martin it takes the light of Arcturus more than one hundred years to reach us. Serviss puts this estimate at forty or fifty years, and states that Arcturus is relatively an aged sun surrounded with a blanket of absorbing metallic vapours, which cut off a large part of his radiant energy, and gives to him a ruddy, fiery hue, especially when he is seen just rising from the horizon. At this time the scintillating colours of Arcturus as viewed in a telescope are beautiful to behold.

It has been proved that we do not receive from Arcturus more heat than we should from a candle at a distance of five or six miles.

Many who were fortunate enough to witness Donati's great comet of 1858 will recall that at one time that comet's head almost occulted Arcturus, and yet its splendour was undiminished.

Prof. Nichols's account of this wonderful sight is worth quoting in this connection:

"It was a spectacle the like of which no one might see again though he should spend on earth fifty lives. At the beginning the comet was like a plume of fire, shaped like a bird of paradise, but it soon brightened into a stupendous scimitar, brandished in the sunset, and when it swept over Arcturus the whole astronomical world was watching to see what would happen to the star."

Arcturus comes to the meridian on June 8th at 9 P.M.

Whitman wrote the following beautiful poem to Arcturus:

Star of resplendent front: thy glorious eye
Shines on me still from out yon clouded sky,
Shines on me through the horrors of a night
More drear than ever fell o'er day so bright,

Shines till the envious Serpent slinks away
And pales and trembles at thy steadfast ray.
Hast thou not stooped from Heaven fair star, to be
So near me in this hour of agony?
So near, so bright, so glorious that I seem
To lie entranced as in some wondrous dream,
All earthly joys forgot, all earthly fears
Purged in the light of thy resplendent sphere,
Kindling within my soul a pure desire
To blend with those its incandescent fire,
To lose my very life in thine, to be
Soul of my soul through all eternity.

The stars β, γ, δ and μ form a trapezium. This figure was known to the Arabs as "the Female Wolves." They also called the star ε Boötis, "Izar" or "Mizar," meaning girdle or waist cloth. This is a double star and its exquisite beauty has earned for it the name "Pulcherima," a title bestowed on it by the elder Struve. The two stars can be seen in a small telescope as the components are 3″ apart.

Canes Venatici
The Hunting Dogs

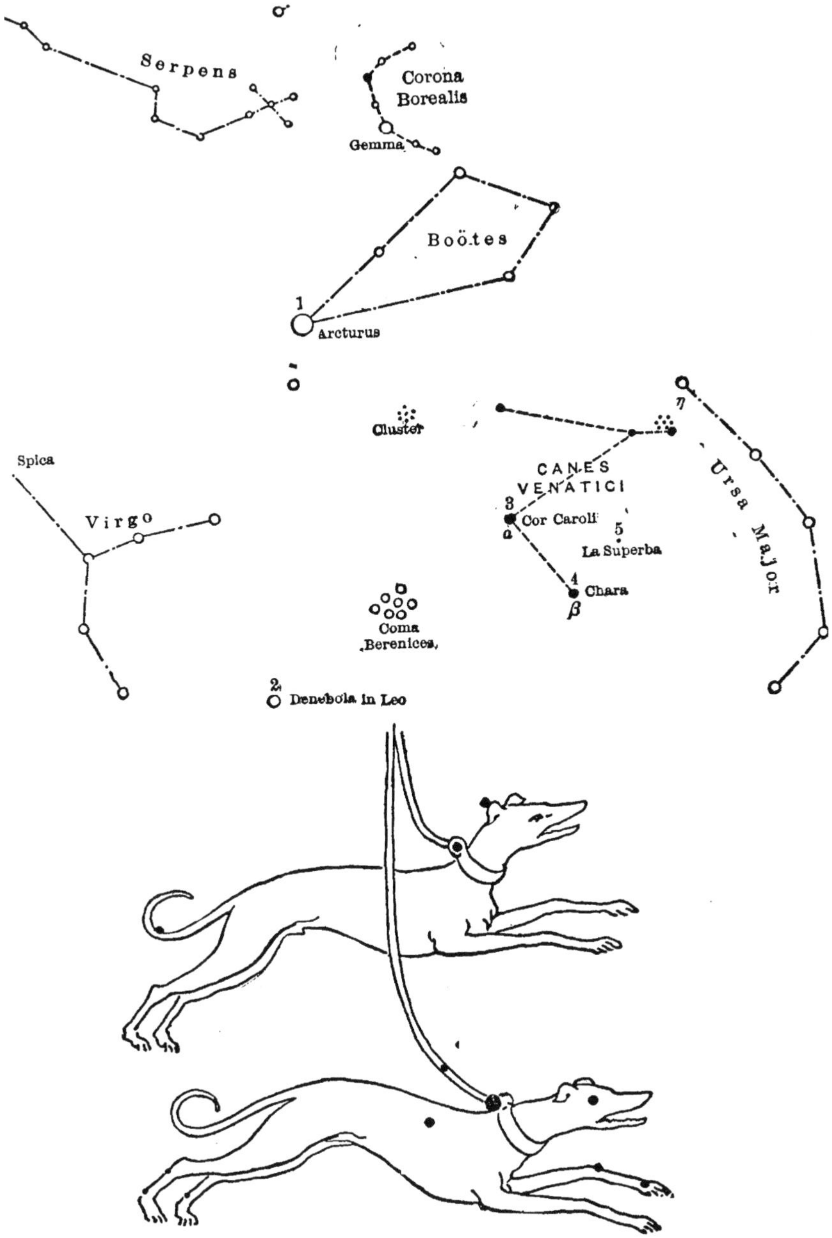

CANES VENATICI

CANES VENATICI
THE HUNTING DOGS

THE asterism Canes Venatici, or the Hunting Dogs, is so closely associated with Boötes that a description of it comes properly in this place.

This star group is a modern one, having been formed by Hevelius in 1690. These stars are supposed to represent two hunting dogs or hounds, which, held in leash by the Bear Driver, pursue the Great Bear as it circles the Pole. The northern dog is named "Asterion," the southern "Chara." In the neck of the latter is situated the lucida of the asterism, a third magnitude star which bears the name of "Cor Caroli," or "Charles's Heart." It was named by Sir Charles Scarborough in memory of Charles I., not Charles II., as often appears. Although it is said Charles II. deserved the honour, as he had the good sense to found Greenwich Observatory. Allen states that this star was set apart in 1725 by Halley, when Astronomer Royal, as the distinct figure Cor Caroli. In China this star was known as "Chang Chen," "a seat," and three stars near the head of Asterion they called "the Three Honorary Guardians of the Heir Apparent." It is a wide double and is easily seen in a small telescope, hence it is a favourite object with amateur astronomers.

Cor Caroli is one of the four stars forming the famous figure known as "the diamond of Virgo," and comes to the meridian at 9 P.M. on the 20th of May.

About seven degrees north and two degrees west of Cor Caroli is the 5.5 magnitude red star which Father Secchi called "La Superba," because of "the superbly flashing brilliancy of its prismatic rays."

The great Spiral Nebula of Lord Rosse, sometimes called "the Whirlpool Nebula," can be seen in this region with a low power about three degrees south-west of η Ursæ Majoris.

Spiral Nebula in Canes Venatici

Cancer
The Crab

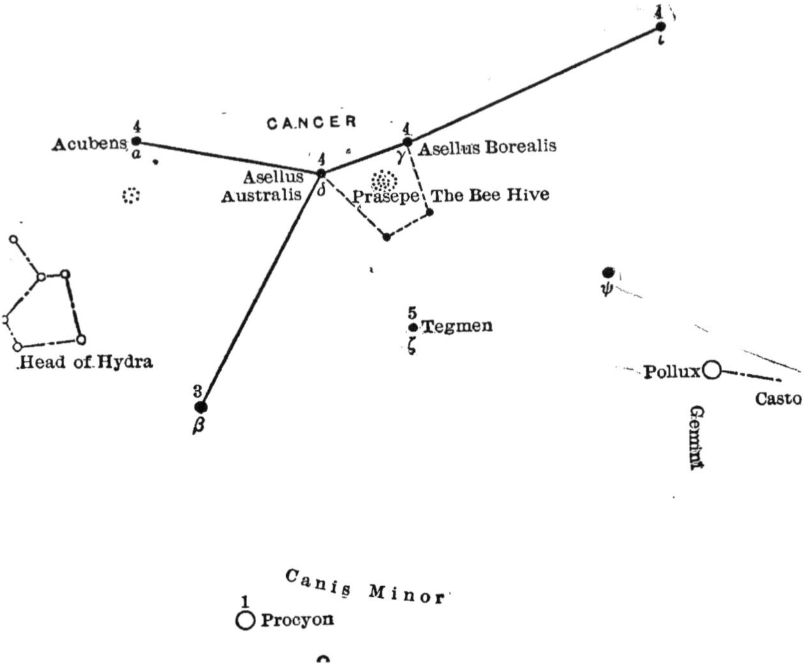
CANCER
Acubens
α
Asellus Australis
δ
Asellus Borealis
γ
ι
Præsepe
The Bee Hive
ψ
Tegmen
ζ
Head of Hydra
β
Pollux
Casto
Canis Minor
Procyon

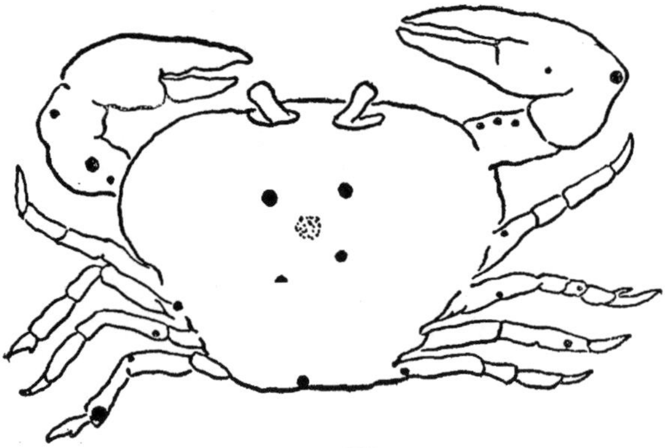

CANCER

CANCER
THE CRAB

The Scorpion's claws here clasp a wide extent,
And here the Crab's in lesser clasps are bent.

CANCER, in spite of the fact that it is the most inconspicuous of all the zodiacal constellations, is very ancient, and has won almost universal recognition in all ages. Because of its dim appearance it has sometimes been called "the Dark Sign," and described as "black and without eyes," and it has been said that among all the constellations not one has been the subject of more idle opinions and more romantic suppositions than Cancer.

Macrobius states that the Chaldeans named the constellation "Cancer" because the crab is an animal that walks backward or obliquely. The sun likewise arriving at this sign begins his apparent retrograde motion and again descends obliquely.

According to Chaldean and Platonic philosophy, "the gate of men," by which souls were supposed to descend into human bodies, was located in this constellation. Plunket tells us that in Babylonia it seems to be established that a tortoise, not a crab, represents the constellation Cancer. It was so figured there and in Egypt in 4000 B.C.

In Egypt, as we learn from the zodiacs of Denderah and Esne, it was the scarabæus, the beetle, emblematic of immortality, that held the place given to the crab in the Grecian Sphere. Burritt thinks that as the Hindus in all probability derived their knowledge of the stars from the Chaldeans, the figure of the crab in this place is more ancient than the beetle.

The crab, tortoise, and beetle, the creatures selected to represent Cancer, are similar in many respects. They are hard shelled, insignificant in appearance, and sluggish in their movements, and in this latter attribute would well typify the sun's apparent movement when it arrives in this constellation.

"If it is admitted," says Plunket, "that in Egyptian astronomy the beetle played the important part of marking as a constellation one of the quarters of the ecliptic circle, then the fact that extraordinary honour is paid in Egyptian symbolic art to this lowly and unattractive insect is explained."

Aratos called the constellation καρκίνος. Latinised it is found as "Carcinus," in the Alphonsine tables. In some Eastern zodiacs Cancer is represented by the figure of two asses, and some of the mediæval astronomers represented it as a lobster or crayfish. In these similes we have, as in the case of the crab, tortoise, and beetle, slow-moving creatures used to represent the constellation, so that there is little doubt that this sign was meant to emphasise the apparent movement of the sun when it was in this part of the zodiac.

According to the Greek legend, this is the crab that seized the foot of Hercules when he was fighting with the Lernean Hydra. The hero crushed the reptile to pieces under his heel, but Juno in gratitude for the offered service, placed the crab in the heavens. Another legend relates that Bacchus, afflicted with insanity, betook himself to the temple of Jove. On the way thither he came to a great marsh, over which he was carried by an ass, one of two which happened to be near at the time. In return for this service, he transformed both creatures into stars. Still another story respecting these stars claims that they owe their place in the heavens to the fact that they were of service to the gods in their battle with the giants. Silenus and Bacchus rode them, and the loud braying of the asses frightened their enemies.

Allen states that Cancer is said to have been the Akkadian "Sun of the South," perhaps from its position at the winter solstice in very remote antiquity, but afterwards it was associated with the fourth month "Duzu" (our June-July), and was known as "the Northern Gate of the Sun." In Yucatan one of the temples was dedicated to Cancer, and the sun when it occupied that sign was supposed to descend at noon like a bird of fire, and consume the sacrifice on the altar.

Cancer is celebrated chiefly because it contains the great naked eye star cluster "Præsepe," the so-called "Manger," from which two asses, represented by stars near by, are supposed to feed. This cluster is known in English astronomical folk-lore as "the Beehive," a name we do not know the origin of. This marvellous aggregation of suns presents on a clear night a dim misty appearance. It has often been mistaken for a comet.

The "Beehive" is especially interesting historically as it afforded Galileo one of the earliest telescopic proofs of the existence of multitudes of stars invisible to the naked eye. He wrote: "The nebula called Præsepe, which is not one star, only, but a mass of more than forty small stars. I have noticed thirty stars besides the Aselli." The great telescopes of the present day reveal in this cluster three hundred and sixty-three stars.

Præsepe has been regarded as representing the Manger in which Christ was born, and Cæsius likened it to the Breastplate of Righteousness. Schiller claimed that Præsepe and the Aselli represented St. John the Evangelist.

The most ancient scientific observation of Jupiter that is known to us was noted by Ptolemy as having occurred eighty-three years after the death of Alexander the Great, when Jupiter happened to pass over the Manger. This was in 240 B.C.

In June, 1895, all the planets except Neptune were in this quarter of the heavens, and here it was that Halley's celebrated comet appeared in 1531.

The Manger was a celebrated weather portent, as early as the days of Aratos and Homer. Aratos thus speaks of it in this connection:

> And watch the Manger like a little mist.
> Far north, in Cancer's territory, it floats,
> Its confines are two faintly glimmering stars,
> One on the north, the other on the south,
> These are two asses that the Manger parts,
> Which suddenly, when all the sky is clear,
> Sometimes quite vanishes, and the two stars
> Seem closer to have moved their sundered orbs.
> No feeble tempest then will soak the leas.
> A murky Manger with both stars
> Unaltered, is a sign of rain.
> If while the Northern Ass is dimmed
> By vaporous shroud, he of the south gleams radiant,
> Expect a south wind. Vapour and radiance
> Exchanging stars, harbinger Boreas.

Pliny wrote: "If Præsepe is not visible in a clear sky it is a presage of a violent storm."

In China the Manger was known, says Allen, by the unsavoury appellation, "Exhalation of Piled-up Corpses," and within one degree of it Mercury was observed from that country on June 9, 118 A.D. One of the Chinese names for Cancer was "the Red Bird," and it was supposed to mark one of the residences of the Red or Southern Emperor.

In astrology, like all clusters, the Beehive threatened mischief and blindness.

In this constellation was located the 6th lunar station of the Hindus, known as "Pushya," meaning "Flower." It was sometimes figured as a crescent, and again as the head of an arrow. If lines are drawn through the stars γ, δ, and θ on the diagram it will be seen that these figures are well named. The Hindu figure of a "flower" in this region of the sky reveals a strange coincidence, to say the least. In Peruvian astronomy Cancer was known as "Cantut Pata," or "Terrace of the Cantut," the cantut being the sacred flower of the Incas. Surely there is more than a coincidence in the fact that two nations, as widely separated as the Hindus and Peruvians, should see in this inconspicuous group of stars a resemblance to a flower. This fact would seem to indicate that at some time in the remote past there was intercommunication between these two great nations.

The cantut flower of the Incas was of a deep red colour, and in June and July the fields around Cuzco in Peru are ruddy with the blooms. The ritual of the Peruvian festival of the sun included the Great Copper Dance, named from the use by the dancers of objects of that dark red metal. At that festival sacred cakes were eaten called "Cancu," made of crushed maize reddened with the blood of animals. The keynote of the ceremonials seems to have been to place emphasis on the colour red, the dark red hidden fire, the colour of the distant but returning sun. The red colour attributed to Cancer accords with the astrological allusion associating Cancer with violent deaths or accidents by fire.

The Arabs knew Cancer as "the mouth and muzzle of the Lion," as to them Leo was a more extensive figure than that known to us, and included Cancer.

The Germans call the constellation "der Krebs,' the French "le Cancri" or "l'Ecrevisse."

The astrological significance of Cancer has generally been malign. It was called "the House of the Moon," from the early belief that our satellite was located in Cancer at the Creation. It governs the breast and stomach, and reigns over Scotland, Holland, Africa, Tunis, Tripoli, Constantinople, and New York. Those born under the sign, that is between June 21st and July 22d, will have a great love of home and family, be quick to feel the mental condition of those around them. Their natures will be quiet and placid, opposed to haste, yet fond of amusement and social pleasures. They dislike quarrels and are slow to change their ideas.

The star Alpha Cancri is a double. It was known to

the Arabs as "Acubens," meaning "the Claws," and marks the Crab's southern claw. It culminates at 9 P.M., March 18th.

The two fourth magnitude stars north and south of the Manger, γ and δ Cancri, were called by the Greeks "the Aselli," the asses feeding at the manger. The Arabs knew them by the same name.

Bailey, in his *Mystic* of 1858, calls them "the Aselline Starlets." The Chaldaic name for the ass may be translated "muddiness," and Burritt thinks that this alludes to the discolouring of the Nile, which river was rising when the sun entered Cancer. Pliny wrote: "Sunt in Signo Cancri duæ stellæ parvæ, aselli appellati." In astrology these stars were portents of violent death to such as came under their influence. They are said to be of a burning nature, and to give great indications of violent and severe accidents by fire.

The star ζ Cancri is a ternary or triple star. Two of the stars can be seen with a small telescope. I quote Allen's reference to this star: "This is a system of great interest to astronomers, from the singular change in colour, the probable existence of a fourth and invisible component, and for the short period of orbital revolution—sixty years—of the two closer stars."

The symbol of the sign ♋ probably denotes the claws of the Crab. It is also referred to the Aselli.

Canis Major
The Greater Dog

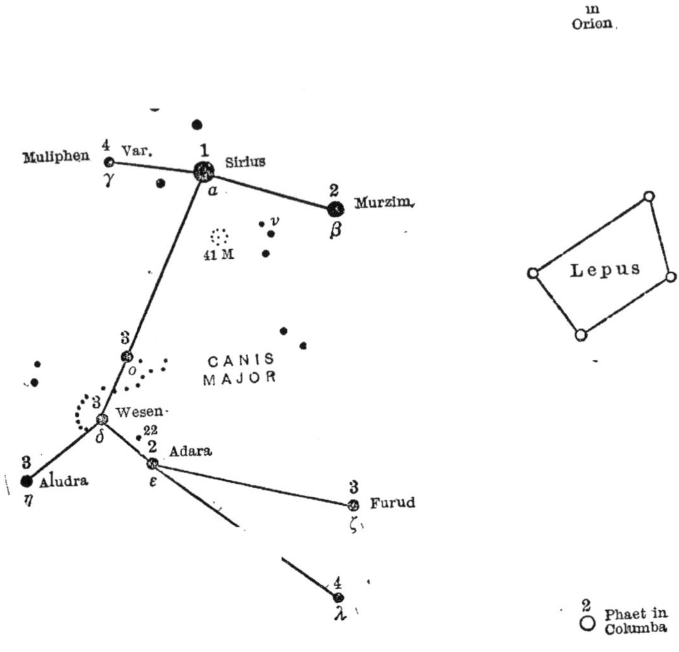

avis

CANIS MAJOR

CANIS MAJOR
THE GREATER DOG

Next shines the Dog with sixty-four distinct;
Fam'd for pre-eminence in envied song,
Theme of Homeric and Virgilian lays.

EUDOSIA.

CANIS MAJOR has been considered from earliest times one of the dogs the giant Orion took with him when he went hunting. Some, however, claim the constellation received its name in honour of the dog given by Aurora to Cephalus, which was the swiftest of his species. The legend relates that Cephalus raced the hound against a fox, which was considered the fleetest of all animals. After they had raced for some time without either obtaining the lead, Jupiter was so much gratified with the fleetness displayed by the dog that he immortalised him by giving him a place among the stars. Another story claims that this was the dog of Icarius.

Among the Scandinavians Canis Major was regarded as the dog of Sigurd, and in ancient India it was called "the Deerslayer." Although mythology connects this star group with the dog of Orion, Sirius, the brightest star in the constellation, seems to have been associated with the idea of a dog even among nations unacquainted with the myth of Orion. In the famous zodiac of Denderah, Canis Major appears in the form of a cow in a boat. It also figures on an ivory disk found on the site of Troy, and on an ancient Etruscan mirror.

According to Burritt, the name and form of the constellation was derived from the Egyptians, who carefully watched its rising, and by it judged of the swelling of the Nile, which they called "Siris."

In the early classical days, says Allen, it was simply "Canis," and represented "Lælaps," the hound of Acteon, or that of Diana's nymph Procris. Homer called it χύων but this doubtless was a reference to the star Sirius. Novidius called it "the Dog of Tobias," and Dr. Seiss regarded Canis Major as "the Appointed Prince."

On the maps, the Dog is generally pictured as standing on his hind feet watching or springing after the Hare, which cowers close under Orion's feet. Bayer and Flamsteed differ from all others in depicting Canis Major as a bulldog. Prof. Young describes "the Greater Dog" as one "who sits up watching his master Orion, but with an eye out for Lepus."

Aratos referring to Canis Major writes: "His body is dark but a star on his jaw sparkles with more life than any other star." This is of course a reference to Sirius, the brightest of all the fixed stars, and probably the star which has attracted the most universal attention of all the heavenly hosts.

In the early histories and inscriptions we find many astronomical references to "the Dog," but it is uncertain whether the constellation or the star Sirius is intended. The Arabian astronomers called the constellation "Al-Kalb-al-Akbar," meaning "the Greater Dog." In the Euphratean star list Canis Major is styled "the Dog of the Sun." Early Christians thought the figure represented Tobias's dog or St. David.

The importance of the constellation is overshadowed by the fame of its lucida, Sirius, the "King of Suns," concerning which star volumes have been written. Its matchless brilliancy has inspired the poets of all ages, and historically Sirius is beyond question the most interesting of all the stars in the firmament.

Aratos thus refers to Sirius:

In his fell jaw
Flames a star above all others with searing beams
Fiercely burning, called by mortals Sirius.

Eudosia writing of the Greater Dog says, "His fierce mouth flames with dreaded Sirius," and Victor Hugo in *The Vanished City* thus alludes to the might of this kingly star:

When like an Emir of tyrannic power
Sirius appears and on the horizon black
Bids countless stars pursue their mighty track.

Aside from the fact of its surpassing brilliance, the fact that Sirius is visible from every habitable portion of the globe has served to make it from time immemorial the nocturnal cynosure of all the nations of the earth.

The sight of this majestic star, clad as it were in all its wealth of history, rising over the snow-crowned hills on a crisp winter's night, flashing to us like a great beacon a message from infinite space, in letters of rainbow hue, is one of entrancing beauty.[1]

The name Sirius is supposed to be derived from the Greek word σείριος which signifies brightness and heat. It is thought by some to represent the three-headed dog Cerberus, who guarded the entrance to Hades, according to Greek mythology.

Allen states that the risings and settings of Sirius were regularly tabulated in Chaldea about 300 B.C., and that it is the only star known to us with absolute certitude in the Egyptian records, its hieroglyph often appearing on the monuments and temple walls throughout the Nile country.

According to Blake,[2] the hieroglyphics representing

[1] Serviss thus mentions Sirius: "The renown of Sirius is as ancient as the human race. There has never been a time or a people in which or by whom it was not worshipped, reverenced, and admired. To the builders of the Egyptian temples and pyramids it was an object as familiar as the sun itself."

[2] *Astronomical Myths*, by J. F. Blake.

Sirius varied in accordance with the different functions the Egyptians ascribed to the star. "When they wished to signify that it opened the year, it was represented as a porter bearing keys, or else they gave it two heads, one of an old man to represent the passing year, and the other of a younger to denote the succeeding year. When they would represent it as giving warning of the inundation of the Nile they painted it as a dog. To illustrate what they were to do when it appeared Anubis had in his arms a stew-pot, wings to his feet, a large feather under his arm, two reptiles, a tortoise and a duck behind him."

Mrs. Martin thus alludes to this glorious sun: "He comes richly dight in many colours, twinkling fast, and changing with each motion from tints of ruby to sapphire and emerald and amethyst. As he rises higher and higher in the sky he gains composure and his beams now sparkle like the most brilliant diamond, not pure white but slightly tinged with iridescence."

Blake gives us the following interesting description of Sirius in the rôle of herald of the inundation of the Nile:

"This star seems to have been intimately connected with Egypt and to have derived its name from that country, and in this way: The overflowing of the Nile was always preceded by an Etesian wind (that is an annual periodic wind answering to the monsoons) which, blowing from north to south about the time of the passage of the sun beneath the stars of the Crab, drove the mists to the south, and accumulated them over the country whence the Nile takes its source, causing abundant rains, and hence the flood. The greatest importance was attached to the foretelling the time of this event, so that the people might be ready with their provisions, and their places of security. The moon was no use for this purpose, but the stars were, for the inundation commenced when the sun was in the stars of the Lion. At this time the stars of the Crab just appeared in the morning, but with them at some distance from the ecliptic, Sirius rose. The morning rising of this star was a sure precursor of the inundation. It seemed to them to be a warning star by whose first appearance they were to be ready to move to safer spots, and thus acted for each family the part of a faithful dog, whence they gave it the name of 'the Dog' or 'Monitor,' in Egyptian 'Anubis,' in Phœnician 'Hannobeach.'"

Sirius, on account of this great service which it rendered the Egyptians, was held in great reverence by them and called "the Nile Star." Under the name Anubis it was deified and this god was emblematically represented by the figure of a man with the head of a dog. It was also worshipped under the names "Sothis" and "Sihor."

Sirius was furthermore known to the Egyptians as "Isis" and "Osiris." If the first letter is omitted from this latter appellation we get "Siris," a name very similar to the modern title of the star.

Other Egyptian names for the star were "Thoth" or "Tayaut" meaning "the Dog," "Hathor," the barker, the monitor, and at Philæ it was called "Sati."

Sirius was worshipped in the valley of the Nile long before Rome had been heard of. In its honour many temples were erected so magnificent in their architectural proportions as to excite wonder and amazement even in this age of noble edifices.

Lockyer found seven Egyptian temples so arranged that the beams from this brilliant star in its rising or setting penetrated to the inner altar, the holy of holies. This feature of architecture is called orientation. Notable among these temples oriented to Sirius was the temple of Isis at Denderah, where Sirius was known as "Her Majesty of Denderah." Here the rising beams of Sirius flashed down the long vista of the massive pylons, and illumined the inner recesses of the temple. What a wonderful sight there must have been enacted within that darkened edifice when, in the presence of a vast multitude silent in meditation, there suddenly appeared a beam of silver light, that laved the marble altar in a refulgence born of the depths of the infinite, a beam, although the watchers knew it not that started on its earthward journey eight and a half years before it greeted their eyes!

The temple priests, versed to some extent in astronomical lore, knew well the psychological moment of the appearance of the light, and doubtless to further increase their prestige, and convey the idea that they were endowed with supernatural powers, so ordered the ritual that the greatest possible superstitious effect would be brought about by the seeming apparition. The awe inspired by the silence of the multitude worked up to a fever pitch of expectancy, and the excitement born of their desire to witness what they must have regarded as a manifestation of divine power, all conduced to make the moment one long to be remembered, and the event one of the greatest possible significance to the race.

It has been determined that the Babylonian star named "Sukudu" or "Kaksidi" was Sirius, for we are told that it was one of the seven most brilliant stars and a star of the south. The same star is also called "directing star" because connected with the beginning of the year.

According to Lockyer, Sirius rose cosmically, or with the sun, in the year 700 B.C. on the Egyptian New Year's Day. In mythological language "she mingled her light with that of her father Ra [the sun] on the great day of the year." This is the first instance of the personification of a star.

"It is possible" says Maunder, "that the two great stars which follow Orion—Sirius and Procyon, known to the ancients generally and to us to-day as 'the Dogs'—were by the Babylonians known as 'the Bow Star' and 'the Lance Star' respectively, the weapons that is to say of Orion or Merodach." Jensen also identifies Sirius with the Bow Star.

Homer compared Sirius to Diomedes' shield, and called it "the Star of Autumn."

Photo by Bonfils

The Temple at Luxor

—— the autumnal star whose brilliant ray
Shines eminent amid the depth of night,
Whom men the dog star of Orion call.

Homer regarded Sirius as a star of ill omen, as it was supposed to produce fevers. Pope's translation of Homer's lines indicates the baleful influence ascribed to Sirius:

A star whose burning breath
Taints the red air with fevers, plagues, and death.

The description of the rising of this star is the only indication in the Homeric poems of the use of a stellar calendar.

Manilius seems to have had two views respecting Sirius. In one place he writes:

All others he excels, no fairer light
Ascends the skies, none sits so clear and bright.

In another we find:

from his [Sirius's] nature flow
The most afflicting powers that rule below.

The Arab name for Sirius was "Al-Shira-al-jamânija," meaning "the bright star of Yemen." Gore thinks that the word "Shira" might have been corrupted in the course of time into Sirius. Al-Shira was also interpreted "the Doorkeeper," Sirius being regarded as the star which opens or shuts. The Arabs also called this star "the Dog Star." In modern Arabia it is "Suhail," the general designation for bright stars.

The so-called "Dog Days" got their name from the fact that in the hottest days of summer Sirius, the Dog Star, blends his piercing rays with those of the god of day. This is of course metaphorical, as the heat we receive from Sirius is inappreciable.

According to Max Müller, the special Indian astronomical name of the Dog Star signified a hunter and deer-slayer.

He is of the opinion that Sirius was called the Dog Star on account of the prevalence of canine madness in the summer season.

Sirius has been appropriately called "the sparkling star" or "Scorcher," and the sun and Sirius have been called "wandering stars." It has been thought that Sirius is identical with the Mazzaroth mentioned in the Book of Job.

It seemed to be the prevailing idea among the ancient Eastern nations that the rising of Sirius would be coincident with a period of heat and pestilence. Virgil well describes the state of affairs when Sirius mingled his beams with those of the day-star.

> Parched was the grass and blighted was the corn,
> Nor 'scape the beasts, for Sirius from on high
> With pestilential heat infests the sky.

Hesiod, who was the first to mention Sirius, wrote in like vein: "When Sirius parches head and knees and the body is dried up by reason of heat, sit in the shade and drink."

Such advice was doubtless as popular then as now during the dog days.

Euripides also refers to the fiery nature of Sirius, describing the star as "sending flames of fire drawn from the heavens."

Apollonius Rhodius speaks of Sirius "burning the islands of Minos."

Horace says: "Here in a quiet valley you will escape the heat of the Dog Star," and in his celebrated ode to the Bandusian Fount he writes:

> 'Gainst flaming Sirius fiery thou art proof.

The question whether Sirius has changed in colour since early times has given rise to considerable controversy. Ptolemy called it fiery red, Seneca claimed it was redder than Mars. Cicero also mentions its ruddy light, and Tennyson wrote:

God Anubis

The fiery Sirius alters hue and bickers into red and emerald.

Dr. See, the eminent astronomer of the present day, asserts that eighteen hundred years ago Sirius was red. There is a reference in Festus to the effect that the Roman farmers sacrificed ruddy or fawn-coloured dogs to save the fruits on account of the Dog Star, and Dr. See says there is no reason why the Romans should sacrifice red dogs except that Sirius was red, and dogs of the same colour must be offered up to the Dog in the sky. There can be no doubt that many of the ancients looked upon red stars as angry deities. Now Sirius is a white star with a bluish tinge, and Allen says that the weight of authority respecting the change in colour of Sirius seems to negative the idea that there has been any change.

Some writers identify the Masonic emblem of the Blazing Star with Sirius, the most splendid and glorious of all the stars.

Topelius, the Finnish poet, fancifully imagines that the great brilliancy of Sirius is due to the combined light of two stars, represented as lovers meeting and embracing:

Straight rushed into each other's arms
And melted into one, ...
So they became the brightest star
In heaven's high arch and dwelt
Great Sirius, the mighty sun,
Beneath Orion's belt.

Although the poet's idea is born of fancy there is nevertheless truth in the statement that we receive from Sirius the combined light of two stars, for Sirius has a faint companion visible only in the most powerful telescopes, and the discovery of this star furnishes an interesting chapter in astronomical history.

The famous German astronomer Bessel expressed his belief about seventy years ago, after ten years of observation, that the periodical variations in the motion of Sirius was produced by the attraction of an invisible companion, revolving around the gigantic star. On Jan. 31, 1862, Alvan G. Clark, at Cambridge, Mass., while testing the 18½″ glass for the Dearborn Observatory at Chicago, pointed the glass at Sirius, when the disturbing companion came suddenly into view at a distance of about 10 seconds from Sirius, and exactly in the direction predicted for that time.

The period of revolution of the companion around Sirius was found to be nearly fifty years, and within a few months of the time calculated by Bessel, long before the telescope had revealed its presence. The mass of Sirius is about twice the mass of its companion, yet its light is 40,000 times greater.

The following facts concerning Sirius may be of interest:

We know now that the brightness of a star is no indication of its distance from us, but Sirius which is 9½ times brighter than a standard first magnitude star is only 8½ light years away, and only four other stars are known to be nearer.

If our sun occupied the place of Sirius in the sky, it would appear as a third magnitude star.

There is considerable discrepancy among the authorities as to the size and brilliance of Sirius as compared with the sun. Its diameter is given as fourteen or eighteen times that of the sun. As regards its brightness, Newcomb states that Sirius is thirty times brighter than the sun, a modest estimate, as other authorities claim for Sirius a brilliance of forty, sixty-three, two hundred, and even three hundred times that of the day-star.

The spectroscope reveals that Sirius is completely enveloped in a dense atmosphere of hydrogen gas. It is the brightest of the so-called Sirian stars, the spectroscopic type I., which includes more than half of all the stars yet studied.

Sirius has a large proper motion—that is the angular change in the position of a star athwart the line of vision—as compared with the average proper motion of stars of the first magnitude. It amounts to 1.31″. In this connection it is interesting to note Al-Sufi's statement concerning the Arab name for Sirius, "Al-abûr." According to this noted astronomer Sirius was so-called because it had passed across the Milky Way into the southern region of the sky. It is a remarkable fact that the proper motion of Sirius would have carried it across the Milky Way from the eastern to the western border in 60,000 years. Possibly the Arabian story may be based on a tradition of Sirius having been seen on the opposite side of the Milky Way by the men of the Stone Age.

It is generally conceded that this star is receding from us, its rate of speed given variously by different authorities as from eighteen to forty miles a second. It comes to the meridian at 9 P.M., Feb. 11th,

The Arab names and meanings of the principal stars in Canis Major, Sirius excepted, are appended:

	Name	*Meaning*
β	Murzim	the Announcer
δ	Wezen	Weight
ε	Adara	the Virgins
ξ	Furud	the Bright Single One
η	Aludra	the Virgin.

Canis Minor
The Lesser Dog

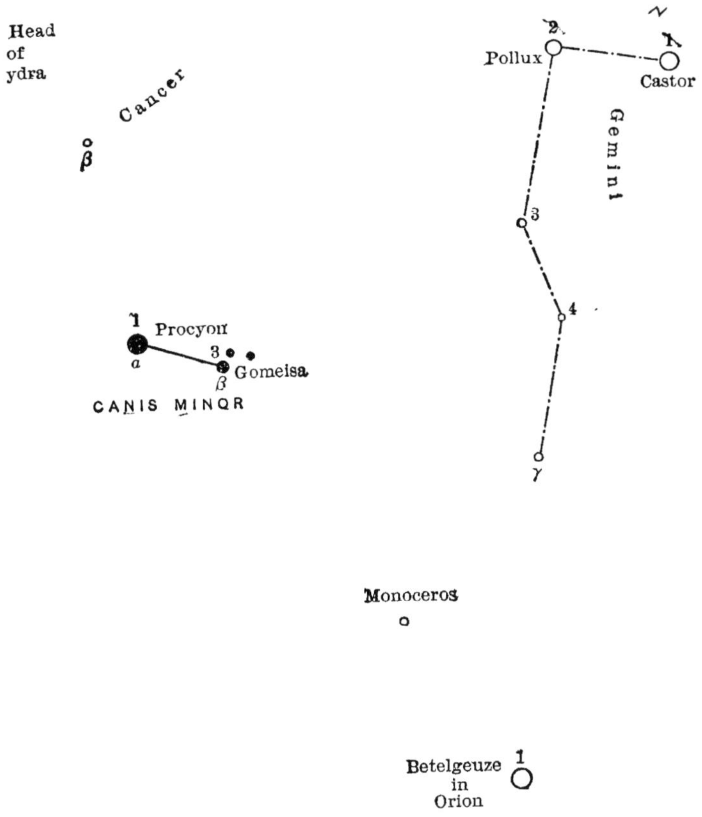
Head
of
ydra
Cancer
β
Pollux
2
1
Castor
Gemini
3
4
γ
1
Procyon
α
3
β
Gomeisa
CANIS MINOR
Monoceros
Betelgeuze
in
Orion
1

CANIS MINOR

CANIS MINOR
THE LESSER DOG

> Canicula, fourteen thy stars; but far
> Above them all, illustrious through the skies,
> Beams Procyon; justly by Greece thus called,
> The bright forerunner of the greater Dog.

CANIS MINOR, according to mythology, was one of the hunting dogs that accompanied the giant hunter Orion, and hence it was sometimes called "Canis Orionis."

Burritt thinks that the Egyptians were the inventors of this constellation, and as it always rises a little before the Dog Star, which at a particular season they so much dreaded, it is properly represented as a little watchful creature, giving notice like a faithful sentinel of the other's approach.

In the valley of the Euphrates it seems to have been regarded as a water dog, on account of its standing on the border of the Milky Way, which represented to the ancients a river in the sky.

Canis Minor has been identified with the Egyptian god Anubis, but Sirius is generally associated with that dog-headed divinity.

Some think the Lesser Dog was the hound of Diana, noted for her love of the chase. Others think that it represents the faithful dog Mæra, which belonged to Icarius, and discovered to his daughter Erigone the place of his burial. It has also been considered to represent Helen's favourite, lost in the Euripus, that she prayed Jove might live again in the sky, and Actæon's hound that devoured his master after Diana had transformed him into a stag.

Schiller thought the figure represented the Paschal Lamb.

The traditional figure of Canis Minor represents it as a well-trained house or watch dog, in contrast with the fierce aspect of the Greater Dog, which is generally depicted as rearing on his hind legs, with the star Sirius blazing in his wide-stretched jaws.

This constellation was included in the great figure of the Lion known to the Arabs, but they called the star Procyon, the lucida of the constellation, "the forerunner of the Greater Dog," and "the blear-eyed Sirius." According to Gore, the Arabs also called Procyon "the Syrian Sirius," because it set in the direction of Syria.

The Romans sometimes called the constellation "Canis" or "Catellus," meaning "the puppy."

Ptolemy accords Canis Minor only two stars, Procyon, and Gomeisa or Gomelza, while Burritt's and Argelander's maps show fourteen and fifteen stars here.

The constellation owes its fame to the first magnitude star Procyon, one of the most interesting stars in the heavens.

"See Procyon too glittering beneath the Twins," says Aratos.

The Greeks called this star προκύων, meaning "before the Dog," the Latin "Antecanis" or " Antecanem," a reference to its rising prior to Sirius. As the rising of Sirius was a warning sign to the Egyptians of the inundation of the Nile, so the appearance of Procyon, the brilliant in the Lesser Dog, warned them still farther in advance of this all-important event. The Babylonians knew Procyon as "the Sceptre of Bel."

In these two constellations of the Greater and Lesser Dogs, we have very good examples of the practical use the stars played in the everyday life of the ancients, and in a measure we see a reason for some of the names of the constellations, which in so many cases seem absurd and irrelevant. Here, as in many of the constellations, there is no resemblance in the configurations of the stars to the figures they are supposed to represent. In Canis Minor

Photo by Brogi

Actæon Attacked by the Hounds of Diana
National Museum, Palermo

with its two stars of any prominence, it would take a fertile imagination to descry the figure of a canine; but when we realise its importance as a warning sign set in the sky for all to observe, then we perceive the significance and appropriateness of the title of the constellation.

Horace, in his celebrated ode to Mæcenas, accredits to Procyon the fiery nature attributed by all to Sirius. He writes:

> Now Procyon flames with fiercest fire;

a line which Mr. Gladstone translates:

> The heavens are hot with Procyon's rays.

Both Sirius and Procyon seem to have conveyed to the ancients the idea of scorching fire and great heat which the dog days at present suggest to us.

Allen tells us that Procyon was "the star of the crossing of the water-dog," mentioned in the Euphratean cylinders and that the natives of the Hervey Islands regarded Procyon as their goddess Vena.

Mrs. Martin referring to Procyon writes: "It is in fact a most beautiful star, and is only the sixth in order of brightness among the stars seen in our latitude. It is very distinctly individual, being the only one among the beautiful winter group that is lightly tinged with yellow. It is one of the Sirian class of stars, but is somewhat further developed than Sirius, and is beginning to have the golden tint which signifies that it is approaching the time of life into which Capella and the sun are well passed."

Al-Sufi, the noted Arabian astronomer, in his *Description of the Fixed Stars*, written in the 10th century A.D., relates the following legend concerning the two Dog Stars: "Al-abûr (Sirius) and Al-gumaïsâ (Procyon) were two sisters of Suhail (Canopus). Canopus married Rigel, and soon after, having killed his wife, fled toward the South Pole, fearing the anger of his sisters. Sirius followed him across the Milky Way, but Procyon remained behind and wept for Suhail till her eyes became weak."

According to Dr. Elkins, Procyon is nine and one half light years from our system, and Vogel claims that it is approaching us at the rate of nearly six miles a second. It is estimated that it emits anywhere from three to eight times as much light as the sun, and it has a thirteenth magnitude companion, discovered in 1896, revolving about it with a period of revolution of about forty years.

Astrologically this star portended wealth, fame, and good fortune. It comes to the meridian at 9 P.M. on the 24th of February.

Beta Canis Minoris is a star of the third magnitude. It was known to the Arabs as "Gomeisa" or "Gomelza" from their name for the constellation, which was "Ghumaisa." This star was noted by Ptolemy, and the Arabs used the distance between this star and Procyon to mark their short cubit, their long cubit being the distance separating Castor and Pollux in the constellation Gemini.

In spite of the fact that Canis Minor is one of the smallest constellations as regards its bounds, it contains four noted variable stars of long period.

Capricornus
The Sea Goat

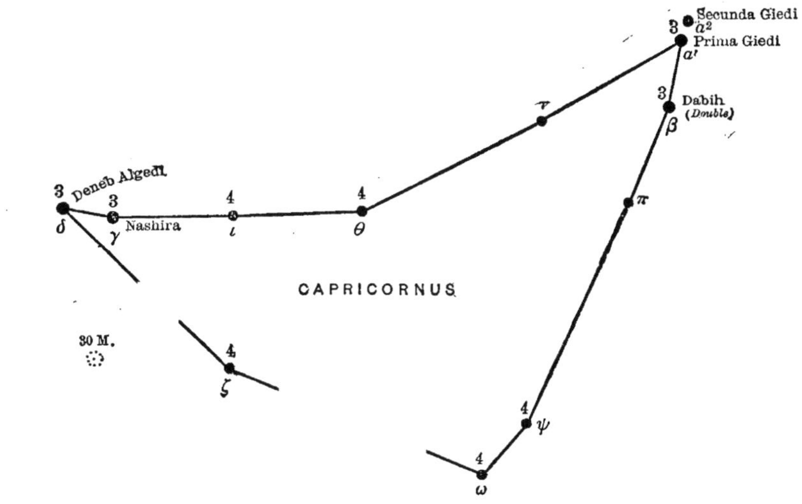

CAPRICORNUS

CAPRICORNUS
THE SEA GOAT

Of Pan we sing, the best of leaders Pan,
That leads the Naiads and the Dryads forth.

BEN JONSON.

VERY few constellations have come down to us unchanged in form through all the ages. An exception to this is found in the figure of Capricornus, which is generally depicted with the head and body of a goat and the tail of a fish.

Allen says that although we do not know when Capricornus came into the zodiac, we may be confident that it was millenniums ago, perhaps in prehistoric days. After Cancer it is the most inconspicuous constellation in the zodiac, and it seems strange on this account that these signs should have held such a place of importance in the minds of the ancients, and that they should have survived without change of figure the assaults of the ages that these stars have gazed upon.

The Capricorn which appears on the Babylonian boundary stones, the most ancient of all records extant, is to all intents and purposes identical in form with the Capricorn of a modern almanac.

According to Macrobius, the Chaldeans named the constellation "the Wild Goat," because that animal in feeding always ascends the hills, and is naturally a climbing animal. The sun in like manner when it arrives at Capricornus begins to mount the sky, and hence the goat was adopted as a symbol of the apparent climbing motion of the sun, while the fish-tail was significant of the rains and floods of the winter season.

This is the explanation of this figure given by most authorities on constellational history to account for the amphibious character of Capricornus. It also explains the ancient oriental legend that Jupiter was suckled by the goat Amalthea, the meaning of which appears to be that the sun, emerging from the stars of Capricornus at the winter solstice, begins to grow in light and heat as he mounts toward the vernal equinox. He is thus figuratively said to be nourished by a goat.

Maunder takes exception to this explanation, and holds that as the constellations were mapped out many centuries before the winter solstice fell in Capricornus, this view of the matter, though ingenious, is illogical and erroneous.

Capricornus was called by the ancient Oriental nations "the Southern Gate of the Sun." In Grecian mythology it was considered "the Gate of the Gods," and through its stars the souls of men released at death were supposed to pass to the hereafter.

Allen tells us that Aratos called this constellation Ἀιγοκέρως the "Horned Goat," to distinguish it from the Ἄιξ of Auriga. The Latinised form, "Ægoceros," was in frequent use with all classical authors who wrote on astronomy.

"The Yoke" was another title borne by the constellation, a name suggested by the configuration of the three principal stars, α, β, and δ. According to Brown, the Akkadai, the most ancient nation known to us, called the tenth month "the cave of the rising" (of the sun), and its nocturnal sign Capricornus, the solar goat, a reduplication of the solar ram, represented the sun rising from the great deep of the under world, as Shakespeare puts it: "from the blind cave of eternal night," and hence a demi-fish.

The Romans considered that Capricornus was under the special protection of Vesta, and they regarded the constellation with great veneration as having shed its influence on the birth of Augustus. We find the figure of a goat on

Typhon
Acropolis Museum, Athens

coins of his period, and Smyth tells us that it was "the very pet of all the constellations with astrologers."

The Arabians also considered Capricornus with great favour, and called it "Al-Jady," meaning "the goat."

Burritt states that Capricornus is identical with Pan or Bacchus, who with some other deities were one day feasting near the bank of the river Nile, when suddenly the dreadful giant Typhon came upon them, and compelled them all to assume a different shape in order to escape his fury. Pan took the lead and plunged into the river, and the part of his body which was under the water assumed the form of a fish, and that above water the form of a goat. To preserve the memory of the fable, Jupiter made Pan into a constellation, in his metamorphosed shape.

The Greeks sometimes called the constellation simply "Pan." From this word we get our word "panic," which is the sort of fear that is born of the imagination, and Pan was said to terrorise people by the mere thought of his presence.

In spite of Pan's evil nature of inciting panics, he was regarded as the god of rural scenery, shepherds, and huntsmen, and also as the god of plenty. The emblem of plenty, the cornucopia or "horn of plenty," is connected with the mythological history of Capricornus.

The legend relates that the father of the gods gave one of the goat's horns to the nymphs who had nursed Jupiter in his infancy as a reward for their kind services, and that this horn was endowed with a wonderful virtue. It provided whatever the holder desired, and hence was known as "the horn of plenty." The real sense of this fable, divested of poetical embellishment, appears to be this: "There was in Crete, some say Lybia, a small territory shaped very much like a bullock's horn, and exceedingly fertile, which the king presented to his daughter Amalthea, whom the poets claim was the nurse of the infant Jupiter" (Burritt).

The emblem of the cornucopia is a masonic emblem, and corroborates the fact that the major part of masonic symbolism has an astronomical significance.

Capricornus is connected in Egyptian astronomy with "the god of waters," and is associated, as the star Sirius is, with the inundation of the Nile. It was also known as the goat-god "Mendes," in the Egyptian zodiac.

Dr. Seiss claims that the Sea Goat represents a symbol of sacrifice and atonement. Cæsius called it "Azazel," "the Scapegoat," and "Simon Zelotis," "the Apostle."

Capricornus marked the 22d Hindu lunar station, "Abhikit," meaning "conquering," and Flammarion asserts that there is a Chinese record of 2449 B.C. which locates among the stars of Capricornus a conjunction of the five planets. There was an early prediction made, that when all the planets met in this sign the world would be destroyed by a great conflagration.

Capricornus has also borne the strange title "the Double Ship," a name that bears out its maritime character appropriately enough, as we find the Sea Goat in that region of the heavens known to the ancients as "the Sea," and surrounded by other creatures of the deep.

Allen states that the symbol of this constellation, ♑, is thought to be τρ, the initial letters of the Greek τράγος, meaning "Goat," but Lalande claims that it represents the twisted tail of the creature. Capricornus figures on an ancient Egyptian mirror. The mirror was emblematic of life, and there may be a connection here between the emblem of life, and the new life established by souls passing through these stars to the life eternal.

The Peruvian year, says Hagar, probably began at the December solstice with the celebration of the most important of their festivals, known as "the festival of the beard." During this month the sun is passing through our sign of Capricornus. The corresponding Peruvian constellation is called "Nuccu," meaning "the

Beard." The name refers directly to the widespread myth in which the sun, then at the height of his power in the southern hemisphere, is figured as Capra, "the bearded one." The beard seems to be the characteristic emphasised in connection with the constellation, and the participants in the ceremonial dances during the festival wore masks with long beards. The beard is one of the chief characteristics of the goat. Thus we find nations widely separated, and at a very remote time, with a common notion respecting an inconspicuous star group. Such a grotesque figure, recognised in common by different nations, is too great a coincidence to savour of individual creation.

It has been said that the tribe of Napthali adopted this sign as their banner emblem, although the sign Virgo has also been allotted to them. The Latin poets designated it as "Neptune's offspring," thus preserving its maritime significance. We also find it called by a Greek appellation signifying "Swordfish," while in the Aztec calendar it appeared with a figure like that of a narwhal. The Tamil name for it signified "Antelope."

Astrologically considered Capricornus was the House of Saturn, the mansion of kings; black russet or a swarthy brown was the colour assigned to it, and Proctor tells us that this sign gives to its natives a dry constitution, and slender build, with a long thin visage. It governs the knees and hams, and reigns over India, Macedonia, Thrace, Greece, Mexico, Saxony, Brandenburg, and Oxford. It is feminine and unfortunate, a conclusion totally at variance with the Romans' exalted idea of the constellation. Those born between the dates Dec. 21st and Jan. 20th are born under this sign. Such persons are proud, self-reliant, and practical, fastidious, dignified, and sincere in affection. Their tendency to idealise brings suffering. March and November are the lucky months, and Saturday the auspicious day. The flower is the snowdrop, and the precious stone, chalcedony.

Aratos thus describes Capricornus:

> the Goat
> Dim in the midst, but four fair stars surround him,
> One pair set close, the other wider parted.

This first pair, a^1 and a^2 Capricorni, respectively called "Prima and Secunda Giedi," are situated in the head of the Sea Goat. Burritt calls them "Giedi" and "Dabih" respectively, the former being the most northern of the two, and a double star. The star name "Dabih" is an Arabic appellation meaning, curiously enough, "the Lucky One of the Slaughterers," referring to the sacrifice celebrated by the Arabs at the heliacal rising of Capricorn.

The other wider parted pair of stars referred to by Aratos are δ and γ Capricorni, named respectively "Deneb Algiedi," meaning "the Tail of the Goat," and "Nashira" —"the Fortunate One" or "Bringer of Good Tidings." δ is an interesting star because it marks the approximate position of the discovery of the planet Neptune.

The discovery of Neptune is one of the most interesting episodes in the history of astronomical discovery, and a brief account of it is worth recording here.

Early in the 19th century it was found that the planet Uranus was straying widely from its predicted positions. Two astronomers, Adams in England, and Le Verrier in France, working independently and without each other's knowledge, endeavoured to ascertain the causes of the perturbations, basing their calculations on the supposition that an undiscovered planet beyond Uranus was the disturbing factor.

Adams began his work in 1843, Le Verrier in 1845. Adams communicated the results of his labour to the Astronomer Royal of England, but unfortunately the data were pigeon-holed. Le Verrier, who sent his calculations to Galle, the eminent German astronomer, was more fortunate. Galle turned his telescope toward the position in the sky determined by Le Verrier, and discovered the planet Neptune. This was on Sept. 23, 1846. Adams at once called attention to his data, which on being referred to were found to coincide with Le Verrier's result. Thus was England robbed of the triumph, but Adams's name has always been coupled with that of Le Verrier as the discoverer of the planet. It may be of interest that the veteran Galle died but a short time ago, July 10, 1910, at the age of ninety-nine.

The remaining stars in the constellation are faint, and of no special interest. When seen on a clear night the constellation resembles an inverted cocked hat.

Cassiopeia
The Lady in the Chair

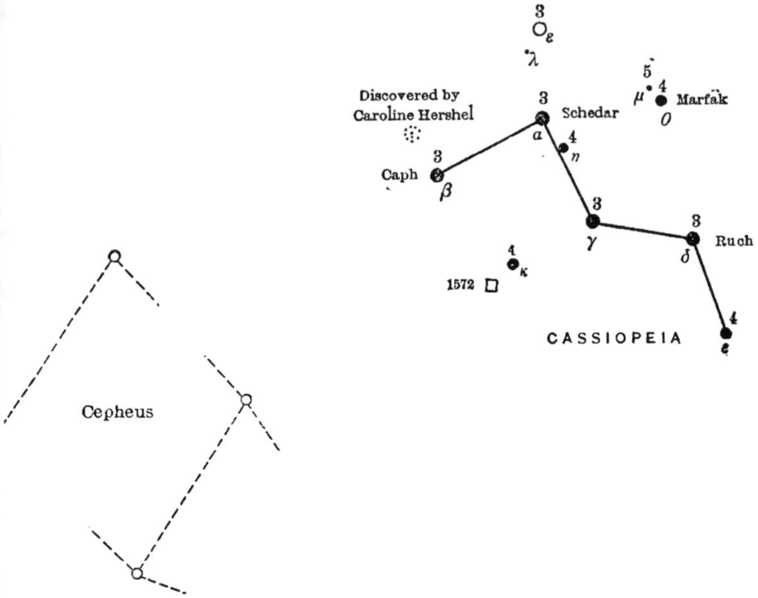
3
ε
λ
5
μ
4
Marfak
o
Discovered by
Caroline Hershel
3
Schedar
α
4
η
3
Caph
β
3
γ
3
Ruch
δ
4
κ
1572
4
ε
CASSIOPEIA
Cepheus

CASSIOPEIA

CASSIOPEIA
THE LADY IN THE CHAIR

. . . look but aside a little,
Just by the first coil of the crooked Dragon
There rolls unhappy, not conspicuous
When the full moon is shining, Cassiopeia.

ARATOS.

CASSIOPEIA is one of the oldest and most popular of the constellations. Popular because many are able to see in the arrangement of its stars the resemblance to a chair, and hence the familiar name for the constellation is "Cassiopeia's Chair."

Such stress has been laid on the throne, that the presence of the Queen seated upon it is lost sight of. Because of the circumpolar motion of the stars, the Queen often suffers the humiliating position of standing on her head. She was placed, so the legend runs, in this cruel position in the heavens by her enemies the sea nymphs, as she had boasted that her beauty surpassed theirs. Desiring to teach her humility they imposed this punishment. Milton in *Il Penseroso* thus refers to Cassiopeia:

that starred Ethiop's queen that strove
To set her beauty's praise above
The sea nymphs and their power offended.

Cassiopeia is sometimes called "heaven troubled queen" and "unhappy Cassiopeia" and in view of the giddy whirl she is subjected to, such appellations are appropriate to say the least. Aratos mentions her uncertain position in the heavens:

> She head foremost like a tumbler sits.

The Arabs called Cassiopeia "the Lady in the Chair," but curiously enough the early Arabs had in this place a very different figure in no way connected with the figure known to us. They called this star group "the large hand stained with henna" or "the tinted hand," the bright stars marking the finger tips. They also made out of the constellations Cepheus and Cassiopeia, two dogs, and some times referred to Cassiopeia as "the kneeling camel."

In this constellation we have, therefore, an example of the fertile imagination of the early Oriental star-gazers, and a curious combination of objects assigned to a group of stars that is not especially conspicuous,—a lady in a chair, a tinted hand, a dog, and a kneeling camel.

As the stars of this constellation revolve about the Pole, they form when below it a slightly distorted capital "M." This is reversed when Cassiopeia is above the Pole, and we have a celestial letter "W" that enables many to identify the constellation.

In Greece at one time this constellation was known as "the Laconian Key," from its fancied resemblance to that article, and Aratos makes the following reference to this title:

> Not many are the stars nor thickly set
> That, ranged in line, mark her whole figure out,
> But like a key that forces back the bolts
> Which kept the double door secured within
> So shaped her stars you singly trace along.

Renouf identified Cassiopeia with the Egyptian star group known as "the Leg," and thus mentioned in the "Book of the Dead," the Bible of Egypt, that most ancient ritual four thousand years old or more: "Hail, leg of the northern sky in the large visible basin."

Cassiopeia belonged to the so-called "Royal Family"

of Starland, and in Greek mythology is connected with the well-known story of Perseus and Andromeda.

Burritt gives the following concise account of the part Cassiopeia played in this drama:

"Cassiopeia was the wife of Cepheus, king of Æthiopia, and mother of Andromeda. She was a queen of matchless beauty, and seemed to be sensible of it, for she even boasted herself fairer than Juno, the sister of Jupiter, or the Nereides, a name given to the sea nymphs. This so provoked the ladies of the sea that they complained to Neptune of the insult, who sent a frightful monster to ravage her coast as a punishment for her insolence. In addition, Neptune demanded a sacrifice of Cassiopeia's daughter Andromeda." The sequel to this sad tale is related in the mythological references to the constellations Perseus and Andromeda.

Brown thinks that this whole story of the sacrifice of Andromeda is Phœnician. He tells us that Cassiopeia was known as "Eurynomê" or "Quassiu-peær," meaning "beauty" or "rosy faced." In the cuneiform inscriptions we meet with the goddess "Kasseba," probably an ancient form of Cassiopeia. On the Assyrian tablets Cassiopeia was "the Lady of Corn," and the Alphonsine tables described the figure as holding the consecrated palm.

There seems to be a decided resemblance between Cassiopeia and the constellation Virgo, which may be nothing more than a coincidence. Virgo we find was called "the Maiden of the Harvest," and was represented as holding a sheaf of wheat or an ear of corn in her hand, and Cassiopeia as we have seen was called "The Lady of Corn."

Again Virgo was represented as a sunburned damsel, while Cassiopeia was called "Æthiop's Queen," clearly indicating her dusky complexion. The Arabs associated dogs with both constellations.

Cassiopeia is represented on some old maps as holding a palm in her left hand. Virgo is invariably represented as carrying a branch in her left hand.

As "the Lady of Corn," Cassiopeia was also designated as "the Creatress of Seed." We also find that the Peruvians identified Virgo with the Earth Mother, and Maunder tells us that the ear of corn in the Virgin's hand may well be interpreted as referring to the "Seed of the Woman" who was born of the Virgin.

Prof. Young has given us a mnemonic word, "Begdi," to assist in recalling the Greek-letter names of the stars in both constellations. In these ways, therefore, there seems to be a distinct similarity between these two female figures widely separated from each other in the starry skies.

Plunket suggests 3500 B.C. as the date, and 23 degrees north as the latitude of the invention of this constellation.

In the 17th century, when there was an effort made to attach a religious significance to the constellations, Cassiopeia became Mary Magdalene, or Deborah sitting in judgment under her palm tree in Mount Ephraim, or Bathsheba, the mother of Solomon, worthy to sit on the royal throne.

The Eskimos imagine that α, β, and γ Cassiopeiæ, three stars forming an isosceles triangle, represent the three stones supporting a celestial stone lamp. They call the constellation "Ibrosi."

Cassiopeia in its continual circling of the Pole of the heavens makes an excellent illuminated timepiece. Imagine that β Cassiopeiæ is the hour hand. When it is above Polaris it is noon, when it is in the west at right angles to its first position, it is 6 P.M. At midnight it is on the northern horizon, and at 6 A.M. it is due east. The time kept by this perpetual clock is of course Sidereal Time (star time), which differs from civil time in that the day begins at noon instead of at midnight. By recalling that the sidereal clock agrees with the mean solar clock on March 22d or thereabouts, and gains at the rate of two hours a month, one can easily pass to ordinary solar time. This is the simplest way to tell time by looking at the stars.

Alpha Cassiopeiæ was known to the Arabs as "Schedar" or "Schedir," meaning "the Breast." Burritt tells us that Schedir is from "El Seder," the "Seder tree," a name given to the constellation by Ulugh Beg. Schedir was discovered to be a variable star by Birt in 1831. It culminates at 9 P.M., Nov. 18th.

Beta Cassiopeiæ, or "Caph" an Arab title meaning "the Hand," was also known to the Arabs as "the Camel's Hump." It is one of the so-called "Three Guides," three stars that mark the equinoctial colure, one of the great circles passing through the poles of the heavens.

Caph is one of the stars for which a parallax has been found. It is approximately twenty light years from our system, though some authorities say thirty-two light years.

Gamma Cassiopeiæ, the second magnitude star in the girdle of the "lady in the chair," has a companion of the 11th magnitude 2″ distant. The Chinese called this star "a whip." It is a star of great interest to astronomers, as it was the first star discovered to contain bright lines in its spectrum. This discovery was made by Secchi in 1886. The spectrum is peculiarly variable.

Delta Cassiopeiæ bears the Arab name "Ruchbah," meaning "the Knee." It was utilised, says Allen, by Picard in France in 1669 in determining latitudes during his measure of an arc of the meridian, the first use of the telescope for geodetic purposes.

Theta and Mu Cassiopeiæ were known to the Arabs as "Al-Marfik," meaning "the Elbow." The star Mu is interesting because of its great proper motion. This is given as 3.7 seconds per year, a velocity in space of one hundred miles a second. It has been estimated that in 3,000,000 years this star will circle the heavens. It is said to be thirty light years distant.

Eta Cassiopeiæ is a double star, and one of the finest objects in the sky for a moderate sized telescope. It is probably the nearest star to us of any in the constellation, although authorities differ as to its parallax. This is given as thirteen, twenty-one, and seventeen light years. The weight of authority seems to favour the latter estimate.[1]

No account of the stars in the constellation Cassiopeia would be complete without a reference to the wonderful temporary star that flashed out in this region of the sky in November, 1572, astonishing the world. It was visible in full daylight, and said to be brighter than the planet Venus. It has been long known as "Tycho's Star," and many conclude from this that it was discovered by the celebrated astronomer Tycho Brahe, but as a matter of fact it was discovered by Schuler, at Wittenberg in Prussia, who saw the star faintly Aug. 6, 1572. Tycho Brahe saw it at its brightest Nov. 11th of the same year, and in 1602 published an account of the star. Other names for this star are "Stranger or Pilgrim Star," "Star in the Chayre," and "New Venus." The Chinese called it "the Guest Star," and Beza thought it was a comet, or the same luminous appearance that guided the Magi, the so-called "Star of Bethlehem."

In March, 1574, the star disappeared entirely. D'Arrest found a minute star of the 10–11th magnitude near this place in 1865 where Argelander could formerly see none. There is some idea that a bright star appeared in this place in the years 945 and 1264 A.D. If so says Webb,[2] we may possibly witness a repetition of this incomprehensible phenomenon.

La Place says: "As to those stars which suddenly shine forth with a very vivid light, and then immediately disappear, it is extremely probable that great conflagrations, produced by extraordinary causes, take place on their surface. This conjecture is confirmed by their change of colour, which is analogous to that presented to us on the

[1] Newcomb writes that β, η, and μ Cassiopeiæ have so great a proper motion in so nearly the same direction that it is difficult to avoid at least a suspicion of some relation between them.

[2] *Celestial Objects for Common Telescopes*, by Rev. T. W. Webb.

earth by those bodies which are set on fire and then gradually extinguished."

Dr. Good thus refers to temporary stars: "Worlds and systems of worlds are not only perpetually creating, but also perpetually disappearing. It is an extraordinary fact, that within the period of the last century, not less than thirteen stars, in different constellations, seem to have totally perished and ten new ones to have been created.

"In many instances it is unquestionable that the stars themselves, the supposed habitation of other kinds or orders of intelligent beings, together with the different planets by which it is probable they were surrounded, have utterly vanished, and the spots which they occupied in the heavens have become blanks."

Burritt thus describes the changes in colour observed in Tycho's star: "At first appearance it was of a dazzling white, then of a reddish yellow, and lastly of an ashy paleness in which its light expired." "It is impossible," says Mrs. Somerville, "to imagine anything more tremendous than a conflagration that could be visible at such a distance." The collision theory seems the best one to account for such phenomena, but the imagination and senses alike fail in any attempt at a realisation of the heat generated by the impact, or the magnitude of the ensuing conflagration.

Cepheus
The King

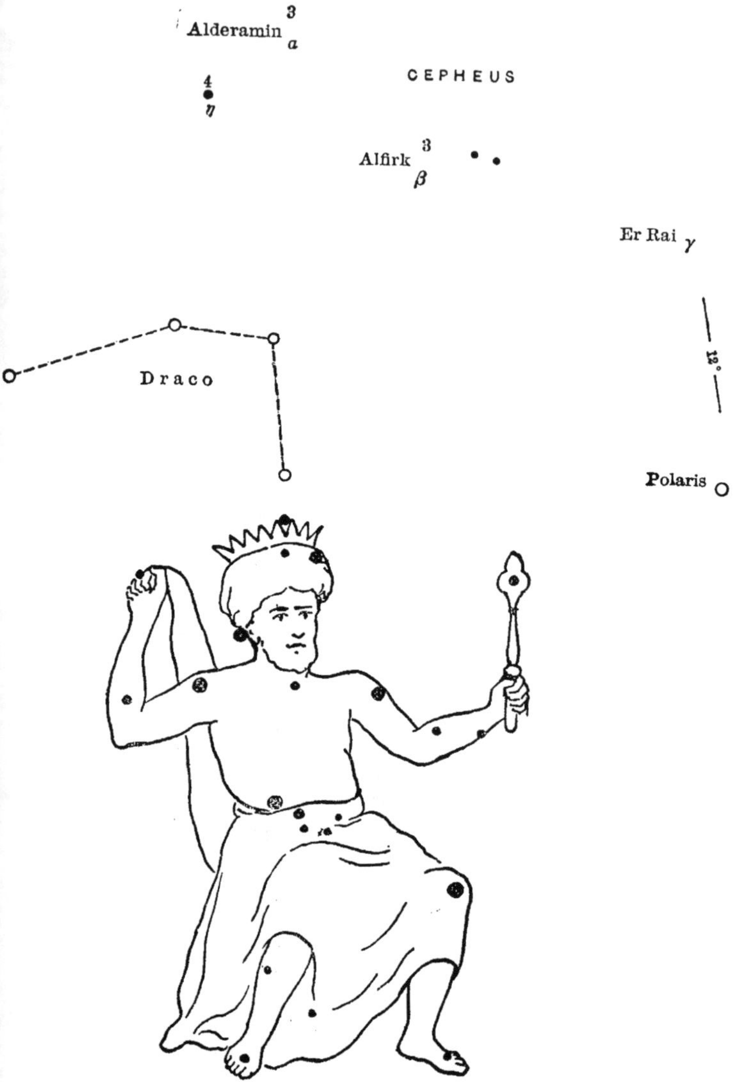

CEPHEUS

CEPHEUS
THE KING

Cepheus himself just behind Cynosura
Stands like one spreading both his arms abroad.

ARATOS.

ALTHOUGH one of the most inconspicuous constellations, Cepheus has attracted attention from the beginning of recorded history. It seems in a measure appropriate that Cepheus should be a dim constellation, for in the thrilling story of the rescue of Andromeda by the champion Perseus, Cepheus, the King, played but a subordinate part.

Plunket gives 3500 B.C. and 23 degrees north latitude as the approximate date and location of the people who invented this constellation. Allen says that Achilles Tatios, probably of our 5th century, claimed that Cepheus was known in Chaldea twenty-three centuries before our era, while according to Brown all of the circumpolar constellations originated on the eastern shores of the Mediterranean.

Cepheus is generally conceded to have been King of Æthiopia, the Euphratean "Cush," the husband of Cassiopeia, and the father of Andromeda. There is a difference of opinion as to what language the names Cepheus and Cassiopeia are derived from. Some writers have suggested for their origin the Sanscrit names "Capuja," which was the later Hindu name for Cepheus, and "Cassyape." Cepheus has also been identified with Cheops or Khufu the builder of the Great Pyramid in Egypt, and, again, was supposed to be descended from Iasion, the son of Zeus and Electra.

Cepheus and the constellations of the group with

which he is generally associated are known as "the Royal Family." They also comprise the so-called circumpolar constellations, and in these latitudes never set. They are especially noteworthy as illustrating the ancient legend of Perseus and Andromeda, one of the best known of all the classic myths and one that has survived all ages. It shows clearly that there was an effort made on the part of the inventor of these constellations to depict here on the imperishable scroll of heaven a drama that should survive all time. There is another such example, which we will come to later, of a like intent to connect a series of constellations, so that the stories that individually relate to each should *in toto* portray a complete history. It is as if each constellation was but an instalment of a serial story. This seems fairly good proof that some of the constellations, at least, were carefully thought out by one man, that design and not chance was responsible for their creation, and that the legends they represented antedated the invention of the several star groups.

Cepheus also figures as one of the Argonauts, the valiant band of heroes that sailed in the ship *Argo* in quest of the golden fleece, and was changed into a constellation at his death. Newton claims that all the ancient constellations relate in some way to this famous expedition. He argues that "as Musæus, one of the Argonauts, was the first Greek who made a celestial sphere, he would naturally delineate on it those figures which had some reference to the expedition. Accordingly, we have on our globes to this day, 'the Golden Ram' (Aries), the ensign of the ship in which Phryxus fled to Colchis, the scene of the Argonautic achievements. We have also the Bull (Taurus) with brazen hoofs tamed by Jason; the Twins (Gemini) Castor and Pollux; two sailors with their mother Leda in the form of a Swan (Cygnus); and Argo, the ship itself. The watchful Dragon (Draco) Hydra, with the Cup (Crater) of Medea, and a raven (Corvus) upon its carcass, as an emblem of death; also Chiron (Sagittarius), the master of

Jason, with his 'Altar' and sacrifice. Hercules, the Argonaut, with his club, his dart (Sagitta), and vulture, with the Dragon, Crab (Cancer), and Lion (Leo) which he slew; and Orpheus, one of the company, with his harp (Lyra). Again we have Orion, the son of Neptune, or as some say the grandson of Minos, with his dogs (Canis Major and Minor), and the Hare (Lepus), River (Eridanus), and Scorpion. We have the story of Perseus, in the constellation of that name, as well as in Cassiopeia, Cepheus, Andromeda, and Cetus; that of Callisto and her son Arcas in Ursa Major; that of Icarius and his daughter Erigone in Boötes and Virgo. Ursa Minor relates to one of the nurses of Jupiter, Auriga to Erichthonius, Ophiuchus to Phorbas, Sagittarius to Crolus, the son of one of the Muses, Capricorn to Pan, and Aquarius to Ganymede. We have also Ariadne's crown (Corona Borealis), Bellerophon's horse (Pegasus), Neptune's dolphin (Delphinus), Ganymede's eagle (Aquila), Jupiter's goat with her kids, the asses of Bacchus (in Cançer), the fishes of Venus and Cupid (Pisces), with their parent the Southern Fish." These, according to Deltoton, comprise the Grecian constellations mentioned by the poet Aratos, and all relate, as Newton supposes, remotely or immediately to the Argonauts.

There is every reason to believe, however, that the constellations were invented long before the date of this famous expedition.

Allen tells us that in China, the Inner Throne of the Five Emperors was located somewhere in this constellation. One of the Chinese Emperors, it is said, ordered a group of stars in Cepheus to be called "Tsau-fu" after his favourite charioteer.

Cepheus had for the Arabs a pastoral significance. In fact in the Euphratean star list Cepheus signified "numerous flock." The stars in the vicinity of the North Pole were supposed to represent a shepherd attended by his dog, watching a herd of sheep at pasture. Goats, calves, and camels also figure in the picture. These animals are all in the neighbourhood of Cepheus.

It is useless of course for us to try to see this picture as it appeared to those night watchers of the far East. Situated in an ideal region for star-gazing as regards climatic conditions, in a land where the nights were glorious with stars and where the people spent most of the nocturnal hours on the house tops or out on the hills, gifted with a wonderfully fertile imagination, it was but natural that they should adore the stars, the mystery of which appealed to their superstitious natures, and exalt their heroes to the starry skies. As they were deeply interested in the care of herds and flocks we naturally find that certain star groups represented to them pastoral scenes. These stellar pictures of the ancients are interesting as showing the changes wrought by the advance of progress and civilisation, and there must indeed have been a fascination in painting pictures on the widespread canvas of the night with a brush steeped in the bright-hued pigments of imagination.

Smyth alluded to the constellation Cepheus as "the Dog," and a ring of stars in this group was known to the Arabs as "a Pot."

Dr. Seiss claimed that Cepheus represented the coming of the Redeemer as King, while Cæsius and Julius Schiller wished to substitute King Solomon and Saint Stephen for the time-honoured personage.

The Cepheid meteor shower of the 28th of June radiates from a point near γ Cephei, and the star μ Cephei is worth observing as being Sir William Herschel's celebrated "Garnet Star," one of the reddest stars in the sky, and a fine object in an opera-glass.

Surrounding the stars δ, ε, ζ, and λ Cephei, which mark the head of the King, is a vacant gap in the Milky Way, one of the so-called "Coal Sacks," where no stars have been observed even in our most powerful telescopes.

Cepheus furnishes a good example of the fact that it is

not always among the brightest constellations that the most interesting objects are found.

Its three brightest stars, α, β, and γ Cephei, gain a certain interest when it is known that by reason of the precession of the equinoxes these stars will one after the other take the place of the Pole Star of ages to come.

In 4500 A.D. γ Cephei will be Polaris. In 6000 A.D. β Cephei succeeds to the title, and 1500 years later α Cephei marks the Pole of the heavens. Only the last will be as near the true Pole as our present Pole star is now.

β Cephei is a beautiful double star, a fine object in a small telescope, and when observing it interest is added by the thought that the primary is also double, although too close to be seen visually, that wonderful instrument the spectroscope revealing its duplicity. This spectroscopic binary has an exceedingly rapid revolution, a complete circuit of the orbit taking less than five hours, which is the most rapid orbital revolution so far known.

The stars ξ and κ Cephei are also fine doubles for a small telescope. For the naked eye observer there is situated in this constellation an object of great interest, the variable star δ Cephei, a typical example of a certain class of variable stars of short period, which are now called the Cepheid variables. Its changes in brightness are perfectly regular and it is an accurate time-keeper, successive maxima following one another at intervals of 5 days, 8 hours, 47 minutes, and 39 seconds. Unlike the so-called Algol variables its light changes are continuous without any period when the brightness is constant. The remarkable behaviour of this star furnishes one of the most puzzling problems of astro-physics. δ Cephei is easily visible to the naked eye and any one can watch these interesting variations in its magnitude. The range is from 3.7 to 4.9.

α, β, and γ Cephei were known respectively by the Arab names "Alderamin," meaning the "right arm," "Alfirk," "a flock," and "Errai," the "shepherd."

Cetus
The Whale

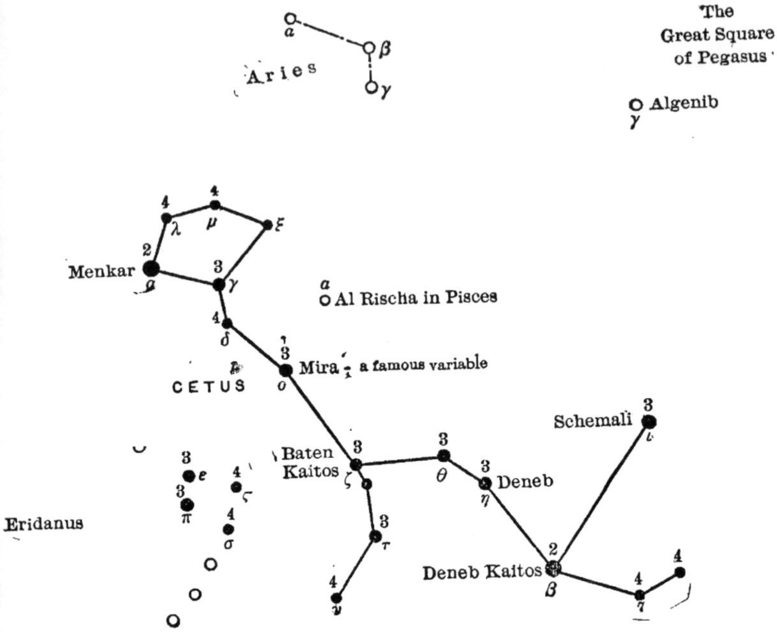

The
Great Square
of Pegasus
Aries
Algenib
Menkar
Al Rischa in Pisces
Mira
a famous variable
CETUS
Schemali
Baten
Kaitos
Deneb
Eridanus
Deneb Kaitos

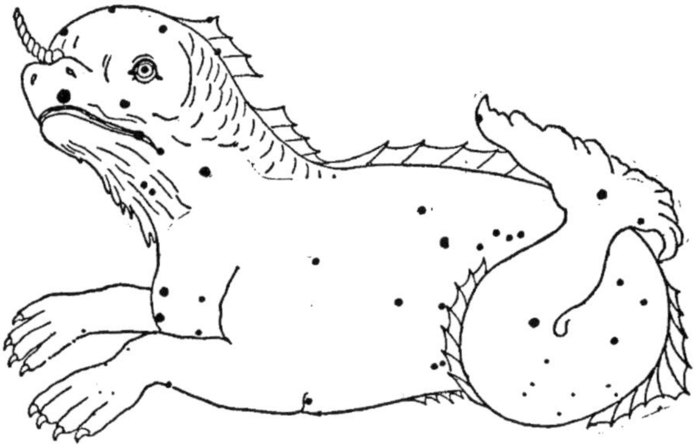

CETUS

CETUS
THE WHALE

With gills pulmonic breathes the enormous whale,
And spouts aquatic columns to the gale;
Sports on the shining wave at noontide hours,
And shifting rainbows crest the rising showers.

DARWIN.

THOUGH Aratos and others connect the Whale with the story of Perseus and Andromeda, there is little doubt that the constellation antedates the time of Perseus.

In earlier times it seems to have been regarded as some kind of leviathan, without connection with the story of the sacrifice of Andromeda. Allen suggests that it may have represented the ferocious Tiamat of the Chaldean myths. In all delineations it has been a strange and fierce marine creature, unlike any known to man, and totally unlike the figure of a whale.

According to Pliny and Solinus, after the monster's encounter with Perseus, in which it suffered from the petrifying gaze of the Medusa, its bones were brought to Rome by Scaurus. Saint Jerome corroborated this story, claiming to have seen the bones of the monster at Tyre.

Brown tells us that Cetus signified "the chaos of the deep" to the Babylonians. It represented primarily the state of chaos "when the earth was waste and wild and darkness was upon the face of the deep." Aratos called it "the dusky monster."

Cetus is sometimes represented as swimming in the river Eridanus, or river Po, the celestial stream into which the venturesome Phaëton was hurled by the bolts of Jove.

Burritt depicts the creature with the two front paws immersed in the River, and the constellation lies between this great stream and the flood which pours forth from the jar of the Water Bearer into the gaping mouth of the Southern Fish. Cetus is thus situated appropriately in that region of the sky known to the ancients as "the Sea," alluded to in a previous chapter, a part of the sky where marine symbols abound.

It has been suggested that these constellations, which might well be designated "the marine group," arranged here together, might have reference to the rainy season, or a period of flood when the sun was in this region of the heavens.

Brown points out the interesting fact that the southern heavens are generally given over to creatures of ill significance. Here we find Hydra, Scorpius, Lupus, Corvus, Canis Major, and Cetus. Design rather than chance seems evident in this arrangement.

In the 17th century Cetus was considered to be a symbol of Jonah's whale, and also of Job's leviathan. Dr. Seiss regards it as the old Serpent, which is the Devil and Satan.

A popular name for this constellation is "the Easy Chair," as the arrangement of its stars suggests to the imaginative a reclining chair. A mutilated hand is also seen by some in the star group forming the head of the creature. The five stars in the head of the whale, α, γ, ξ, μ, and λ, form a fairly regular pentagon, which serves as a ready means of identifying the constellation.

The arrangement of the stars in Cetus permits of many geometrical figures being formed. The stars ζ, θ, τ, η, and ι Ceti form an inverted dipper, a little larger but otherwise not unlike the so-called "Milk Dipper" in the constellation Sagittarius.

The body of the creature is kite-shaped, and the entire constellation somewhat resembles the figure of the prehistoric ichthyosaurus.

Although Cetus is the largest constellation, it contains

few telescopic objects of interest. The south pole of the Milky Way is located within its borders, and the constellation "is a condensation point of nebulæ, directly across the sphere from Virgo, also noted in this respect."

Alpha Ceti is no longer the lucida of the constellation, as its Greek-letter name would indicate, for it is inferior in brightness to Beta. One or both of these stars have therefore changed in the course of time. Alpha is well worth observing as a fine combination of a beautiful 2.5 magnitude orange-coloured star with a 5.5 magnitude star of a decided bluish tint. The Arab name for Alpha is "Menkar," meaning "the nose." λ Ceti also bears this name, and as it is situated exactly in the nose of the creature it seems more appropriately named than Alpha. Astrologically Menkar denoted sickness, disgrace, and ill fortune, with danger from great beasts.

β Ceti was known to the Arabs as "Diphda" or "Deneb Kaitos." Diphda signifies "the Frog," and this star was called "the Second Frog," the first one being represented by the star Fomalhaut situated in the mouth of the Southern Fish. The name "Deneb Kaitos" means "the Tail of the Whale toward the South." In China this star bore the strange title of "Superintendent of Earthworks."

No account of the constellation Cetus would be complete without a reference to the wonderful variable star Mira, or Omicron Ceti as it is generally called by astronomers. Historically it is the most interesting of all the variable stars of long period, and it bears the distinction of being the first star whose variability was discovered.

D. Fabricius observed it early on the morning of the 12th of August, 1596, as somewhat brighter than α Arietis. In October it had disappeared. He observed it again in February and March, 1609. Holwarda observed it in 1638, and recognised its periodical variability.

According to Argelander's calculations its period is 331½ days, but it is very irregular, and the difference of period is sometimes as much as twenty-five days. Its

10

magnitude at maximum also varies greatly. At times it vies with stars of the second magnitude, and often it only attains a brilliance of the fifth magnitude. At minimum it is generally of the ninth magnitude, only a thousandth part of its greatest brilliance, and one twentieth as bright as the faintest stars visible to the naked eye.

Mira is of a deep red colour and gives an interesting spectrum. Espin points out a similarity between the spectrum of Mira and the celebrated Nova Aurigæ. Herschel notes an observation of Omicron Ceti on the 6th of November, 1779, when this wonderful star equalled Aldebaran in brightness.

The amateur astronomer with a telescope of 3″ aperture or better can observe very well all the changes in light that take place in this remarkable star. Change seems to bespeak life, and hence the observance of variable stars must ever prove a source of fascination and wonder to those who make a study of them, for they, of all the seeming lifeless orbs that gaze so steadfastly on the centuries, exhibit inherent qualities that distinguish them in the firmament as man is distinguished on earth.

For three centuries this star has been under observation and as yet shows no sign of relaxation.

No satisfactory theory has yet been found to account for all the variations in the light of these long-period variables. It has been suggested that the irregularities are caused by the phases of some general law, like the law of the maxima and minima of sun spot activity.

τ Ceti is one of our nearest neighbours in space, its distance being estimated as nine light years.

Corona Borealis
The Northern Crown

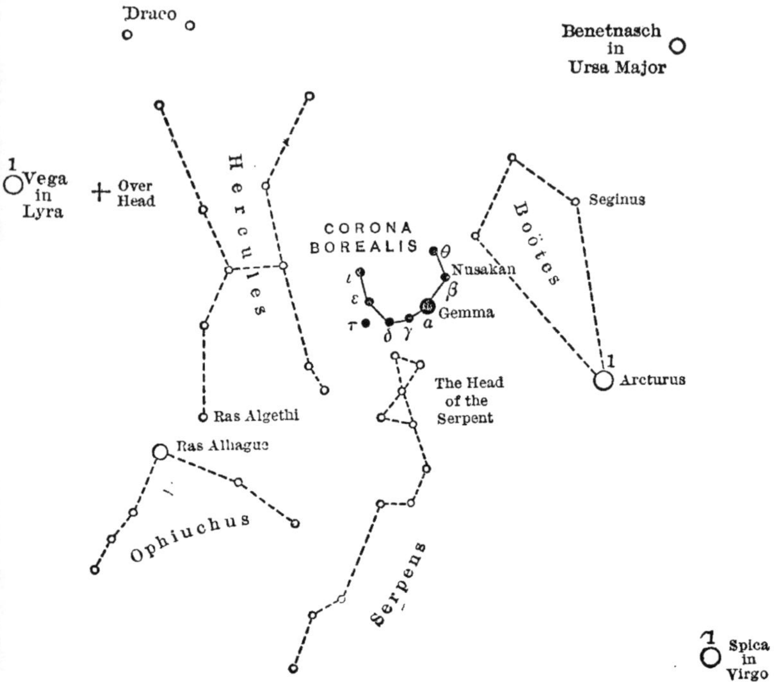

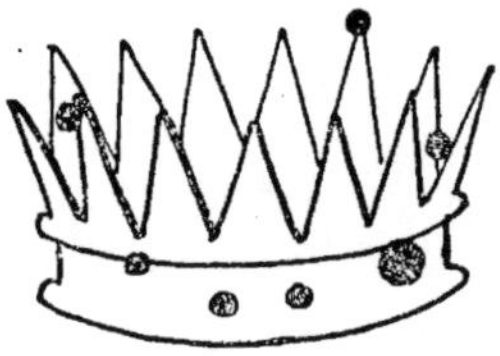

CORONA BOREALIS

Theseus Slaying the Minotaur
Statue at Villa Albani

CORONA BOREALIS
THE NORTHERN CROWN

There too that Crown which Bacchus set on high,
A brilliant sign of the lost Ariadne.

ARATOS.

THIS conspicuous and beautiful constellation is said to commemorate the crown presented by Bacchus to Ariadne, the daughter of Minos, second King of Crete. The legend relates that Theseus, King of Athens (1235 B.C.), was shut up in the celebrated labyrinth of Crete to be devoured by the ferocious Minotaur, which was confined in that place. This creature was accustomed to feed upon the chosen young men and maidens exacted from the Athenians as a yearly tribute to the tyranny of Minos. Theseus attacked and slew the wicked monster, and being furnished with a clue of thread by Ariadne, who was passionately devoted to him, he extricated himself from the difficult windings of the labyrinth. He afterwards married the beautiful Ariadne, and carried her away to the island of Naxos, where sad to relate he deserted her.

Ariadne was so disconsolate at this treatment, that some say she hanged herself, but Plutarch takes a more cheerful view, and claims that she lived many years after and was espoused to Bacchus, who loved her with much tenderness and gave her a crown of seven stars, which after her death was placed among the stars. Thus the constellation is often called "Ariadne's Crown."

Spenser however thinks that Theseus was the donor of the crown. In his *Faerie Queen* he says:

Look: how the crowne which Ariadne wore
Upon her yvory forehead . . .
Being now placed in the firmament,
Through the bright heavens doth her beams display,
And is unto the starres an ornament,
Which round about her move in order excellent.

Apollonius Rhodius thus refers to the Crown in his *Tale of the Argonauts* as early as the third century B.C.

Still her sign is seen in heaven,
And midst the glittering symbols of the sky
The starry crown of Ariadne glides.

Brown claims that the crown was bestowed by the sun-god Dionysos on his consort Ariadne (the very chaste one) on the occasion of his nuptials in the island of Naxos.

We therefore have our choice as to who bestowed the crown on Ariadne—Bacchus, Theseus, or Dionysos.

Allen tells us that Pherecydes, in the fifth century before Christ, was the first to record this legend of Ariadne's Crown, and the constellation is without doubt one of great antiquity. It is one of the few that resemble in the arrangement of stars relative to each other the subject supposed to be represented. The stars are arranged in a semi-circle, and outline a perfect crown, so that this group is easily identified, and because of its beauty is better known than many of the constellations.

This constellation has also been regarded as "the Coiled Hair of Ariadne," a reduplication of the asterism Coma Berenices or Berenice's Hair.

One of the most peculiar features of the arrangement of the stars into constellations by the ancients is the fact that many of the figures are repeated, and in almost every case the two constellations similar in figure are situated close together in the sky. Thus we find two Dogs, two Lions, two Bears, two Birds, two Giants (Hercules and Ophiuchus), two Fishes, two Crowns (the northern and southern), two Centaurs, and now as we have seen above

The Minotaur
Painting by George Frederick Watts

there seem to have been two constellations that represented maiden's tresses, only separated by the constellation Boötes.

This fact of reduplication seems to corrobrate the evidence that there was a deliberate plan exercised in the designing of the constellations, for there were many animals known to the ancients that are not given a place in the stellar menagerie. There must have been some very good and sufficient reason for duplicating so many of the star groups.

The Northern Crown has also borne the following titles: "The Wreath of Flowers," "Diadema Cœli" "Oculus," meaning any celestial luminary, and "Mæra," signifying the "shining one."

The fact that the stars forming the Crown do not form a complete circle has caused it to appear other than crown-like to various peoples. Thus it is said to resemble a Beggar's Dish with a nicked rim, such as is held out by the beggar to receive alms.

The Australian natives called this constellation "womera," our boomerang, the arrangement of the stars suggesting that weapon to their minds.

The Shawnee Indians of our own country called this constellation "the Celestial Sisters," and have an interesting legend respecting it, which is a typical example of the imaginative power possessed alike by the red men of North America, and the far-off nomadic tribes of the ancient world. The legend is as follows: "White Hawk, a mighty hunter, was searching for game. He suddenly found himself on the outskirts of a great prairie, where he perceived a circular path worn through the grass with no path leading to it. While he stood wondering at the strange pathway, he saw descending from the heavens a silver basket containing twelve beautiful maidens. As the basket touched the ground they alighted and began dancing about the ring, beating time on a silver ball. White Hawk endeavoured to capture the most beautiful of the maidens, but they all leaped into the basket which was instantly carried up into the sky. The next day White Hawk revisited the spot disguised as a rabbit, and tried in vain to seize one of the dancers. The day following in the guise of a mouse he was more successful, and succeeded in catching the most bewitching maiden, and took her home as his bride. She soon became homesick, however, and one day when White Hawk was absent she made a silver basket, and singing her magic chant was carried to the heavens, where she appears now as one of the bright stars near the Crown, the star Arcturus in the constellation Boötes."

The Indians also imagined that this star-traced circle represented a council of Chiefs, and the star in the centre of the circle was the servant, cooking over the fire, preparing the feast.

Manilius in the first book of his *Astronomicon* thus speaks of the Crown:

> Near to Boötes the bright Crown is viewed,
> And shines with stars of different magnitude.

Corona Borealis was known to the Hebrews by the name of "Ataroth," and by this name the constellation is called in the East to this day.

Cæsius said that this Crown represented the one that Ahasuerus placed upon Esther's head, or the golden crown of the Ammonite King, of a talent's weight. He also likened it to the Crown of Thorns worn by the Christ.

This constellation is especially interesting as marking the region of the sky where the most celebrated temporary star of recent years appeared. It was observed 58′ south of the star Epsilon, on the 12th of May, 1866, as a second magnitude star, and was visible for eight days. The star then slowly declined to the tenth magnitude and rose later to the eighth. Now it appears a pale yellow, and is known as T Coronæ. It is slightly variable. This was the first temporary star to be studied by the spectroscope.

The brightest star in the constellation is Alpha, a

Photo by Mansell

star of 2.4 magnitude. It was known to the Arabs as "Alphecca," which means the "bright one of the dish," referring to the resemblance of the constellation to a broken plate mentioned above. This star is also called "Gemma," and "the Pearl of the Crown," a title which Allen says has been occasionally transformed into Saint Marguerite.

Gemma is receding from our system at the rate of about twenty miles a second. Manilius thus refers to it:

> One placed in front above the rest displays
> A vigorous light and darts surprising rays.
> This shone since Theseus first his faith betray'd,
> The monument of the forsaken maid.

This star marks the radiant point of the Coronids, a meteor shower visible from April 12th to June 30th. It culminates at 9 P.M., June 28th.

The Northern Crown contains the exceedingly interesting variable star lettered "R," which has been called "Variabilis Coronæ." It was discovered by Pigott in 1795 and varies with much irregularity from 5.8 to the 13th magnitude. There are only two other variables known to be of this type. To give an idea of the rapid changes of light in this star, it is interesting to note that on Sept. 21, 1910, its magnitude was 6.5, Oct. 17th it had declined to a 9.6 magnitude star, and by Nov. 2d it was only a dim star of the 12th magnitude.

The Greek word βαγδει is the mnemonic word given by Young to assist the memory in locating the stars in this constellation, the stars in the Crown bearing in sequence these Greek-letter names.

Photo by Anderson

Ariadne Sleeping
National Museum, Rome

Corvus
The Crow

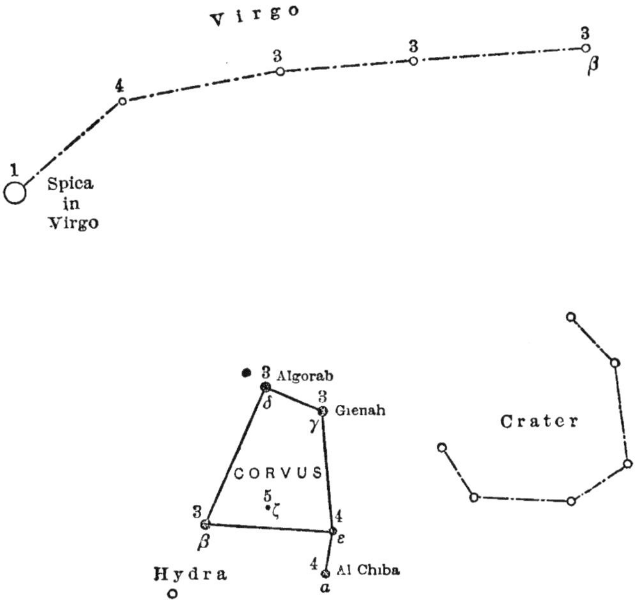
Virgo
3
3
3
β
4
1
Spica in Virgo
3 Algorab
δ
3 Gienah
γ
Crater
CORVUS
5
ζ
3
β
4
ε
4 Al Chiba
α
Hydra

CORVUS

CORVUS
THE CROW

> The figure of a crow seems pecking at him.
>
> ARATOS, referring to Hydra.

ON most of the ancient star maps, the Crow is generally depicted as perched on the coils of the great water snake Hydra, and apparently "pecking at him," as the poet puts it.

The ancient Akkadians, according to some authorities, seem to have regarded this constellation as representing a horse, but nearly all the other ancient nations saw in this group of stars a bird.

With the Chinese it was "the Red Bird," the last constellation in their zodiac. The Romans and Hebrews called this constellation "the Raven," the name it was known by in Chaucer's time, and Brown tells us that in the valley of the Euphrates there was a connection between Tiamat, the Serpent of Night, and the Demon Ravens. It was known there as "the Great Storm Bird," "the Bird of the Desert," "the Bird of the Great Seed," and "Storm Wind."

It is said that the crow was once of the purest white, but was changed to his present sable hue for talebearing.

Thus is the fact immortalised in verse:

> The raven once in snowy plumes was drest,
> White as the whitest dove's unsullied breast,
> Fair as the guardian of the capitol,
> Soft as the swan, a large and lovely fowl;
> His tongue, his prating tongue, had changed him quite,
> To sooty blackness from the purest white.

According to the Greek fable, the crow was made a

constellation by Apollo. This god being jealous of Coronis (whom he loved), the daughter of Phlegyas, and mother of Æsculapius (who is represented in the skies by the figure of the giant Ophiuchus), sent a crow to watch her behaviour. The bird perceived her criminal partiality for Ischys, the Thessalian, and immediately acquainted Apollo with the fact, which so fired his indignation that:

> His silver bow and feather'd shafts he took,
> And lodged an arrow in her tender breast,
> That had so often to his own been prest.

To reward the crow he placed it among the constellations, but just why it was located on the back of the Hydra does not appear.

It is a curious fact that we find elsewhere among the constellations birds closely associated with other figures. Thus the Crane is shown as pecking at the Southern Fish. The Pleiades or Doves flock together on the back of the ferocious Bull, and ancient Chinese maps depict the Eagle on the Dolphin's back. Design clearly enters into these grotesque arrangements.

Some say that this constellation takes its name from the daughter of Coronæus, King of Phocis, who was transformed into a crow by Minerva, to rescue the maid from the pursuit of Neptune.

Allen gives the following classical legend respecting Corvus: It appears that the bird being sent by a god with a cup for water, loitered at a fig tree till the fruit became ripe, and then returned to the god with a water snake in his claws, and a lie in his mouth, alleging the snake to have been the cause of his delay. In punishment he was for ever fixed in the sky with the Cup and the Snake, and, we may infer, doomed to everlasting thirst by the guardianship of the Hydra over the Cup and its contents. Hence the constellation has been called "Avis Ficarius," the "Fig Bird," and "Emansor," one who stays beyond his time. There is a belief in early folk-lore that the crow alone among birds does not carry water to its young.

Corvus, Crater, and Hydra are generally associated together in the ancient myths and legends. Swartz, early in the 19th century, endeavoured to prove that the constellations were nothing but a sort of symbolical geography of the west shore of the Caspian Sea. He imagined that these three constellations represented strangely enough the petroleum wells of Baku. The long extended Serpent, with its coils and folds, represented to him the slow, oily flow of crude petroleum. The Cup is placed there to indicate the receptacle or reservoir for the oil, and the Crow is indicative of the inky blackness of the colour of the oil.

Dr. Seiss regards the Crow as the Bird of Doom, and it has been likened to Noah's Raven flying over the waste of waters, or alighting on Hydra, as there was no dry land for a resting place.

This association of the Crow with the bird that went forth from the Ark connects this constellation with several others that many authorities believe form a graphic account of the Deluge, and, just as in the Perseus and Andromeda group we seem to find a serial story, here is depicted in a like group the story of the Flood.

It certainly seems plausible that primitive man should have sought to record the greatest and most important events known to him for the benefit of posterity, and it may have occurred to some ancient patriarch, and possibly Noah, to inscribe his record on the enduring scroll of night, and burn the legend deep with the fire of the silver stars.

At any rate, there is a significant arrangement of constellations in this region of the heavens, that requires little imagination to convey a fairly good record of the Deluge story, as we have it in Genesis.

Here we have the Ship (the Ark), Argo, stranded upon a rock. Two birds hover near-by, the Raven and the Dove, the birds sent forth by Noah. We have a sacrifice offered up by a person who has gone forth from the Ark, the Centaur, and we see in the sky the Altar, and smoke arising from it represented by the Milky Way. Curiously enough we also find a Bow set in a cloud, not the rainbow, but the Bow of the Archer, set in the Milky Way, the cloud of smoke. This connection of Sagittarius with the group of Deluge pictures may seem a bit far fetched, but even without it the picture of the Flood and the story in Genesis are well borne out in the constellations, and we find in this group the best of evidence that they were combined and placed here as a record for all time.

In addition to the constellations named as belonging to this group, Aquarius and Eridanus have been said to represent the Deluge, and Pisces and Cetus, the fishes and whale swimming in the "deep waters."

In Genesis, as Maunder points out, Noah is represented as a man. In the constellation picture, he who issues forth from the Ship is a Centaur, one who partook of two natures. There is certainly a significance in these figures of the Centaurs. They were a very ancient people, regardless of the fact whether such creatures ever existed, or whether, as has been supposed, they were people who tamed horses, and appearing on horseback, an uncommon sight, resembled at a distance a figure half man, half horse.

The significance of these figures in the heavens lies in the fact that the inventor of the constellations was familiar with the figure of a horse, which we find depicted in Pegasus and the distinct figure of the ancient Centaur, half man, half horse.

Corvus was known to the Arabs as "the Camel," and "the Tent." It forms the 11th Hindu lunar station, known as "Hasta," meaning the "Hand." Schickard thought Corvus represented Elias's Crow.

The Arabs called Alpha Corvi "Al-Chiba," which was also the Desert title for the constellation. Ulugh Beg and other Arabian astronomers called it "the Raven's Beak."

Delta Corvi, called by the Arabs "Algorab," is a beautiful double star, a fine object for a small telescope, the colour contrast, yellow and purple, being especially pronounced.

Crater
The Cup

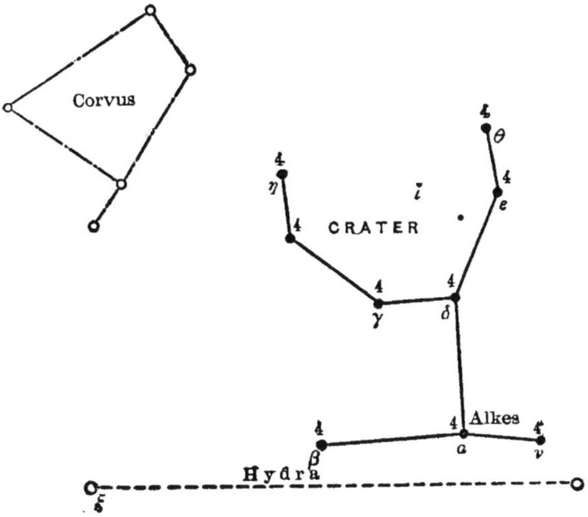

CRATER

CRATER
THE CUP

Midway
His volume is the Cup.
ARATOS, referring to Hydra.

CLOSELY identified with Corvus is the constellation Crater, the Cup, an inconspicuous group of stars bounding Corvus on the west. An imaginary line, drawn through the brighter stars in Crater, traces out a bowl-shaped figure, whence its fancied resemblance to a Cup, the title the constellation has borne from time immemorial.

In the old atlases, the Cup is usually represented in the form of a large urn elaborately ornamented, with two handles set opposite each other and rising above the rim of the bowl, resting insecurely on the coils of the great sea serpent Hydra.

This was the cup fabled to belong to Bacchus, and Manilius thus refers to it:

the generous Bowl
Of Bacchus flows and cheers the thirsty Pole.

The original connection of Crater and Corvus is with Hydra, the storm and ocean monster. Crater was the symbol of the vault of heaven, wherein at times storm winds, clouds, and rain were chaotically mixed, while Corvus, as we have seen, was known as "the Great Storm Bird."

Omar in the following familiar lines employs the simile respecting Crater and the dome of heaven:

And that inverted Bowl they call the sky,
Whereunder crawling coop'd we live and die.

Earlier in the *Rubáiyát* we find the Cup and the Bird mentioned in one quatrain:

> Come fill the Cup and in the fire of Spring
> Your winter garments of Repentance fling:
> The Bird of Time has but a little way
> To flutter—and the Bird is on the wing.

It is possible that no astronomical significance was intended here by this reference to the Bird and Cup, as the Bird is clearly Time and not the Raven, but the Bird and Cup were so closely identified in the astronomical lore of the Orient, that the Persian poet may well have considered the simile more fitting than would at first appear. It is certainly a curious coincidence.

Allen tells us that in the early Greek days Crater represented the κάνθαρος or "Goblet of Apollo," which universally was called κρατηρ. The Greeks also called the Cup κάλπη, a "cinerary urn," and ʽγδρία, a "water bucket."

One Greek legend connected Crater (the Mixing Bowl) with the Cup of Icarius, to whom Bacchus gave the wine, and who was translated to the sky as the constellation Boötes. Another, originating in Asia Minor, connected the Cup with the mixing of human blood with wine in a bowl.

In China the constellation figured strangely enough as a dog.

In the Euphratean star list, Crater is called "the Bowl of the Snake." Other names for it are the Cup of Herakles, of Achilles, of Dido, of Medea.

No allusions to this constellation have as yet been found in the excavated relics of ancient Egypt, although Allen informs us that there is an ancient vase in the Warwick collection on which is the following inscription:

> Wise ancients knew when Crater rose to sight,
> Nile's fertile deluge had attained its height.

Photo by Brogi

Medea

National Museum, Naples

There certainly would seem to be a significance attached to Crater in Egyptian star lore, as Hydra, so intimately connected with Crater, has been regarded as the inhabitant of the Nile, and in fact its representative. In all probability, evidence connecting this constellation with Egyptian astronomy will come to light in the near future, as the work of excavation is rapidly going on in that rich land of buried treasure.

In early Arabia this constellation was known as "the Stall," a figure much resembling the Manger in the constellation Cancer. Hewitt connects the Cup with the Soma Cup of prehistoric India. It has also been identified with the cup that Joseph found in Benjamin's sack, with Noah's wine cup, and the cup of Christ's Passion. Dr. Seiss regarded it as the Cup of Wrath of the Revelations.[1]

The constellation contains no stars of special interest.

Inasmuch as Crater was regarded in ancient times as the symbol of the vault of heaven, it may be well to remark here an interesting fact, often lost sight of, concerning the stars in their relation to our planet, which a recent writer has pointed out.

To the individual, the heavens resemble nothing so much as "the inverted Bowl" of which Omar sings, with its rim resting on the hills, and other irregular surface features that limit our view of the horizon, but it is obvious that the sky is much more extended, as it covers half the earth, and is not bounded by the individual's horizon.

Therefore, when we look at the stars, some of them are twinkling above the billows of the mighty oceans, the Atlantic, the Pacific, the Indian. Others look down on the lofty Andes, with their snow peaks, and pierce the gloom of the tropical forest in the valley of the Amazon. Still other suns glitter above the bergs and field ice of the frozen polar seas, on scenes of frigid desolation, and some of the

[1] Early Christians believed that the constellations Corvus and Crater represented the Ark of the Covenant.

stars we nightly gaze upon are mingling their beams with the arc lights of our great and populous cities.

"Every point in the sky is directly above some point on the earth, and as is the proportion of a given area to the whole sky, so is the proportion of the area it overhangs to the whole surface of the earth." Thus the Pleiades, the famous cluster in Taurus, cover a space about equal to Westchester County, N. Y.

Cygnus
The Swan
or
The Northern Cross

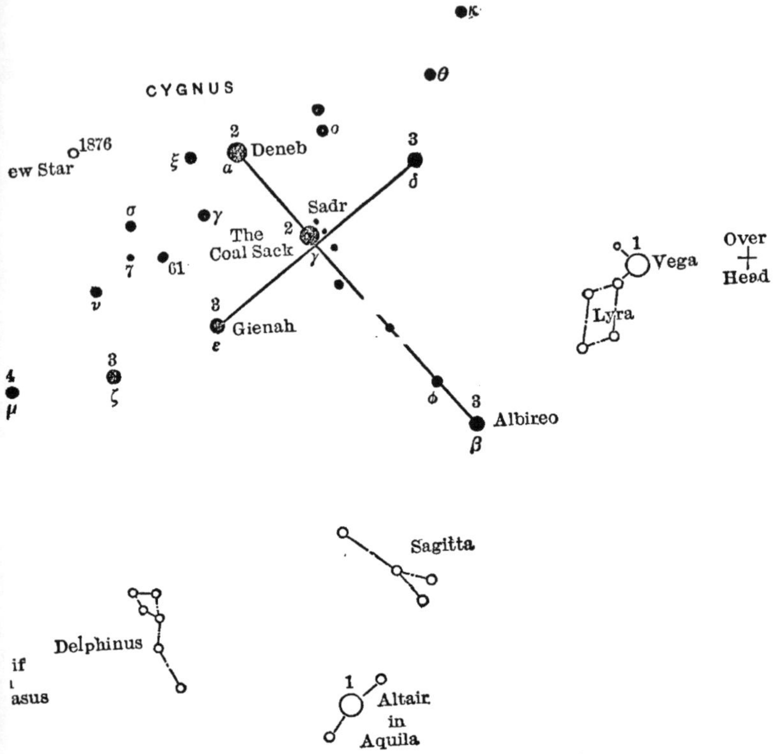
CYGNUS
1876
ew Star
Deneb
Sadr
The Coal Sack
Gienah
Albireo
Vega
Over Head
Lyra
Sagitta
Delphinus
if
asus
Altair in Aquila

CYGNUS

CYGNUS
THE SWAN
OR
THE NORTHERN CROSS

Thee, silver Swan, who silent, can o'erpass?
A hundred with seven radiant stars compose.
The graceful form: amid the lucid stream
Of the fair Milky Way.

EUDOSIA.

THERE are few constellations in the firmament that exceed in beauty and interest the star group popularly known as "the Northern Cross." Shrouded in the glory of the Galaxy, rich in telescopic objects of exquisite beauty, famed in fable and song, this constellation possesses a charm that is enduring, a fascination for all students of the stars and lovers of the beautiful.

There are various legends to account for the presence of the Swan in the starry skies. One relates that the Swan represents Orpheus the wonderful musician, who won the beautiful Eurydice for his bride. Foully slain by the cruel priestess of Bacchus, Orpheus was changed into a swan and transported to the heavens, where he was placed near his beloved Harp (the constellation Lyra), possibly to add his mite to the sweet music of the spheres.

Others suppose it to be the swan into which Jupiter transformed himself when he deceived Leda, wife of Tyndarus, King of Sparta.

Again we are told that the Swan was Cicnus or Cycnus (and this is believed to be the proper name for the constellation), a son of Neptune, who was invulnerable to attack from blows or missiles. Achilles after striving in vain to wound Cicnus finally succeeded in smothering him. As he was about to rob his victim of his armour, Cicnus was suddenly changed into a swan.

According to Ovid, the constellation took its name from Cygnus, a relative of Phaëton's, who deeply lamented the untimely fate of that youth, who was hurled into the river Eridanus after his disastrous ride. The legend relates that after Phaëton had disappeared beneath the waters of the river, Cygnus frequently plunged into the stream to seek him. The gods in wrath changed him into a swan, and therefore it is that the swan ever sails about in the most pensive manner, and frequently thrusts its head beneath the water.

Virgil in the 10th Book of his *Æneid* thus alludes to this fable:

> For Cicnus loved unhappy Phaëton,
> And sung his loss in poplar groves alone,
> Beneath the sister shades to soothe his grief.
> Heaven heard his song and hastened his relief,
> And changed to snowy plumes his hoary hair,
> And winged his flight to sing aloft in air.

Allen tells us that this constellation may have originated on the Euphrates, for the tablets show a stellar bird of some kind. At all events the present figure did not originate with the Greeks, for the history of the constellation had been entirely lost to them.

In Arabia Cygnus was called "the Flying Eagle," and "the Hen," appearing under the latter title about 300 B.C. in Egypt.

Cygnus is generally represented in full flight along the Milky Way.

> Yonder goes Cygnus the Swan, flying southward.

On some old maps the bird is apparently just rising from the ground. Aratos describes the Swan:

> As one that floats on well poised wings.

Photo by Annan

Orpheus and Eurydice
Painting by George Frederick Watts

In the Euphratean star list this constellation bears the title of "Bird of the Forest."

Before the time of Eratosthenes (the third century B.C.) the name of the star group among the Greeks was simply "the Bird."

This portion of the sky seems to abound with birds,—as if they hovered over the region of the heavens known to the ancients as "the Sea." Here we find besides the Swan, Aquila the Eagle, the flying Eagle, and Lyra, which was regarded as the falling or swooping Eagle.

Sayce states that the Assyrian name of the Swan is supposed to be "Tussu," while Houghton has been unable to discover any Hebrew, Assyrian, or Phœnician name for the constellation.

The brightest stars in Cygnus form the so-called "Northern Cross," a perfect and beautiful figure, which Lowell thus alludes to in his poem, "New Year's Eve, 1844":

> and countless splendours more
> Crowned by the blazing Cross high hung o'er all.

The Cross is formed by the stars α, γ, η, β Cygni, marking the upright along the Milky Way, more than twenty degrees in length, and ζ, ε, γ, δ Cygni, forming the transverse.

The early Christians regarded this figure as the Cross of Calvary, as did Schiller. The Northern Cross is certainly much more perfect in form than the famed Southern Cross, and setting in the west, when it assumes an upright position, it presents a beautiful appearance.

Christmas eve at nine o'clock, this brilliant cross of stars stands upright on the western hills, outlined against the sky as if beckoning all beholders onward and upward. A beautiful symbol of the Christian faith, glorious, perfect, and eternal, and especially significant at this season of the year.

Between α, γ, and ε Cygni is one of the vacant spaces in the Milky Way, a black and seemingly bottomless abyss, the brink over which man peers into the profound and mysterious depths of interstellar space. This wonderful region in the sky is known as "the Northern Coal Sack."

Cygnus is celebrated as containing the nearest lucid star in the northern hemisphere, a 5.6 magnitude star, barely visible to the naked eye, known as "61 Cygni." It is a double star, both being golden yellow in colour, situated six degrees north and east of the star ε Cygni. Bessel, in 1838, calculated its distance as approximately six light years,—a light year being the distance light travels in a year at the rate of 186,000 miles a second. 61 Cygni bears the distinction of being the first star whose distance was measured. If the distance from the earth to the sun equals one inch, the distance from the earth to 61 Cygni would equal seven and one half miles. This gives one a good idea of star distances, which to all intents and purposes are beyond our comprehension. To illustrate the amount of labour bestowed by astronomers on the problem of the determination of star distances, it may be mentioned that for the photographic measure of the star 61 Cygni, 330 separate plates were taken in the year 1886–7. On these, thirty thousand measurements were made. The result agreed closely with the best previous determination by Sir Robt. Ball, using the micrometer.

The lucida of the constellation is Alpha Cygni, called by the Arabs "Deneb," meaning the "Hen's Tail." It is a brilliant white 1.4 magnitude star, comparatively young, and in the same spectroscopic class as Spica and Vega. The spectroscope reveals that Deneb is approaching the earth at the rate of thirty miles a second. From the best determinations Deneb is at least ten times as far off as Vega. The distance of Vega is thus expressed: Supposing the sun's distance from the earth (93,000,000 miles) equals one foot, Vega is 158 miles distant. This puts Deneb at a reasonably safe distance from us.[1]

[1] Deneb is also in Newcomb's "XM" class of stars, embracing the stars that are thousands or perhaps even hundreds of thousands of times brighter than the sun.

Deneb can be seen at some time between sunset and midnight every night in the year in these latitudes. Mrs. Martin pays it this charming tribute: "Deneb is particularly attractive in the early evening in January and February. It is then rather low in the north-west and with Vega gone, and no other bright star very near, it has a more commanding charm. In the heavier atmosphere which induces a more rapid twinkling, the star seems to take on an accession of gaiety, and goes fairly dancing down behind the horizon, where it finishes its circle and appears again in the north-east about four hours later."

Deneb culminates at 9 P.M., Sept. 16th.

Cygnus is a splendid field for telescopes, great and small. Its chief object of beauty is the incomparably beautiful double star, Beta Cygni, also known as "Albireo," situated in the base of the Cross and in the beak of the Swan. Even in a small telescope the contrast in the colours of these two close set stars is well emphasised, and the sight of these suns, the one gold, the other blue, never fails to charm all who view them. As this double is easily split by small telescopes, Albireo is a great favourite with all amateur astronomers.

Cygnus contains many deeply coloured red and orange stars, and Birmingham called this part of the heavens "the Red Region," or "the Red Region of Cygnus."

Espin gives a list of one hundred stars in this constellation that are double, triple, or multiple.

The 2.7 magnitude star γ Cygni was called "Sadr" by the Arabs, meaning the "Hen's Breast." Allen says, "it lies in the midst of a beautiful stream of small stars, itself being involved in a diffused nebulosity extending to α Cygni, while the space from γ to β Cygni is perhaps richer than any of similar extent in the heavens." In this space, according to Herschel, the stars in the Milky Way seem to be clustering into two separate divisions, each division containing more than 165,000 stars. So rich is this region of the heavens in stars, that Herschel counted 331,000 in a width of only five degrees.

In the neck of the Swan, not far from Beta Cygni, is the variable star χ Cygni, discovered by Kirch in 1686, ranging from magnitudes 4.5 to 13.5 in 406 days. Sometimes at its maximum it reaches only the sixth magnitude, thus presenting a problem which is one of the most interesting in all the realm of astrophysics.

Delphinus
The Dolphin

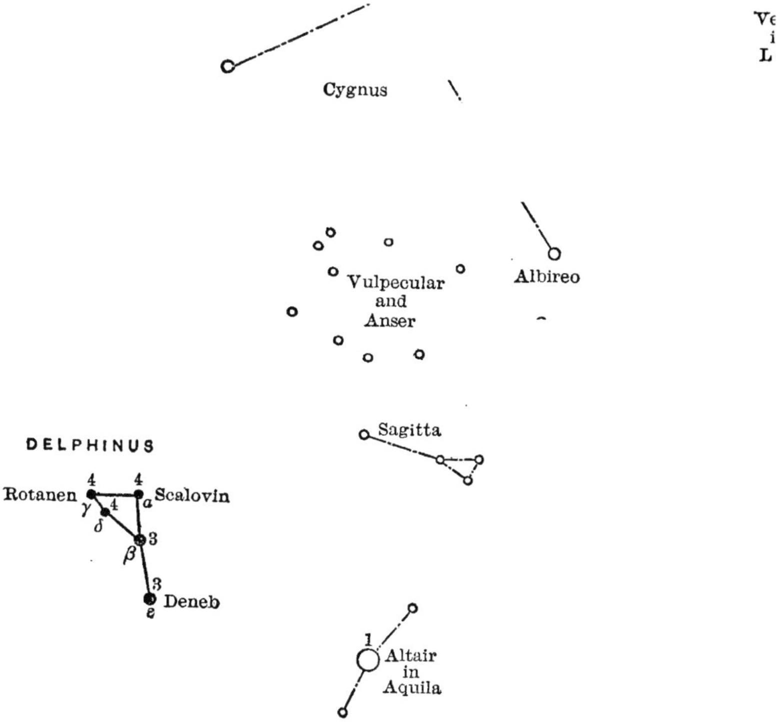
Cygnus
Albireo
Vulpecular
and
Anser
Sagitta
DELPHINUS
4
4
Rotanen
Scalovin
γ
4
α
δ
3
β
3
Deneb
ε
1
Altair
in
Aquila

DELPHINUS

DELPHINUS
THE DOLPHIN

The Dolphin small to sight floats o'er the Goat.
ARATOS.

THIS interesting little constellation lies in that region of the sky already alluded to as "the Sea," and near-by are the other maritime creatures, the Fishes, the Sea Goat, and the Whale.

In Ptolemy's catalogue Delphinus only contained ten stars; Burritt gives it eighteen, Argelander twenty, Heis thirty-one,—none brighter than the third magnitude. Allen thinks that the constellation originally may have included the stars set off by Hipparchus to form the asterism Equuleus.

In all astronomical literature, Delphinus has borne its present title and shape, but just why the Dolphin should be represented by these stars is not clear.

A favourite title for the constellation was "Vector Arionis," from the Greek fable which Burritt thus relates:

"The Dolphin was made a constellation by Neptune, because one of these beautiful fishes had persuaded the goddess Amphitrite, who had made a vow of perpetual celibacy, to become the wife of that deity;—but others maintain that it is the dolphin which preserved the famous lyric poet and musician Arion, who was a native of Lesbos, an island in the Archipelago. Arion went to Italy with Periander, tyrant of Corinth, where he obtained immense riches by his profession. Wishing to revisit his native country, the sailors of the ship in which he embarked resolved to murder him, and get possession of his wealth. Seeing them immovable in their resolution, Arion begged permission to play a tune upon his lute before he should be put to death. The melody of the instrument attracted a number of dolphins around the ship. He immediately precipitated himself into the sea, when one of them, it is asserted, carried him safe on his back to Tænarus, a promontory of Laconia, in Peloponnesus, whence he hastened to the court of Periander, who ordered all the sailors to be crucified at their return."

> But (past belief) a dolphin's arched back
> Preserved Arion from his destined wrack;
> Secure he sits and with harmonious strains
> Requites his bearer for his friendly pains.

Spenser pays the following tribute to the friendly dolphins:

> "Then was there heard a most celestial sound
> Of dainty music which did next ensue,
> And, on the floating waters as enthroned,
> Arion with his harp unto him drew
> The ears and hearts of all that goodly crew;
> Even when as yet the dolphin which him bore
> Through the Ægean seas from pirates' view,
> Stood still, by him astonished at his lore;
> And all the raging seas for joy forgot to roar.

The dolphin is also said to have performed a friendly service in the cause of Justice. Hesiod, the famous poet, having been slain and his body cast into the sea, the dolphins recovered the body and conveyed it to the shore. Here it was found by his friends, who hunted down the assassins aided by the poet's dogs, and put them to death by drowning in the sea into which they had thrown Hesiod.

A curious coincidence is revealed by this legend, for here we find the dolphin identified with the preservation of a corpse, and the constellation is popularly known as "Job's Coffin." There can hardly be any connection between these similar allusions, as in all probability the title "Job's

Photo by Brogi

Cupid and Dolphin
National Museum, Naples

Coffin" was applied long after the constellation was known as "the Dolphin."

Allen says that he has been unable to learn the date and name of the inventor of the title "Job's Coffin." The stars in the constellation form a rectangular figure not unlike the shape of a coffin, which possibly accounts for the title.

We find the dolphin acting as a life-saver again in the rescue of Taras, the founder of Tarentum, in Italy, from a watery grave. The inhabitants of that city struck a coin in memory of this event.

Delphinus marked the 24th Hindu lunar station, known as "Most famous." In Greece Delphinus was the Sacred Fish, the creature being of as much religious significance there as the fish afterwards became among the early Christians. The Dolphin was also regarded as the messenger and favourite of Poseidon, and the sky emblem of philanthropy.

The Arabs called this constellation "the Riding Camel," and the early Christians are said to have believed that this star group represented the Cross of Jesus, transferred to the sky.

The dolphin has also been regarded as the fish which swallowed Jonah, although this title, properly speaking, should apply to Cetus the Whale. Schiller knew some of the stars in Delphinus as "the water pots of Cana." The Chinese called the four principal stars in that constellation, which form a diamond-shaped figure, "a gourd."

Delphinus in astrology was believed to have a special influence over the births and character of human beings.

It has been thought that this constellation was invented by a seafaring people, and this, with the neighbouring star groups, is evidence that the constellations were in all probability designed by dwellers on the coast.

Some regard the dolphin as one of the many animals worshipped in connection with Apollo. Again it is said to be the symbol of spring, and the opening of the season of navigation, and others claim that the title is derived from the name Delphi, as the festival known as "the Delphinia" was celebrated in May, and commemorated "the genial influence of the spring sun on the waters, in opening navigation, and in restoring life to the creatures of the wave, especially to the dolphins which were highly esteemed by the superstitious seafaring fishermen, merchants, etc."[1]

The star names "Sualocin" and "Rotanev," applied to α and β Delphini, are interesting as presenting a mystery for many years. Webb finally discovered that reversing the spelling gave "Nicolaus Venator," the Latinised name of the assistant to the astronomer Piazzi, who is thus immortalised in starland.

The star γ Delphini is a fine double star for a small telescope, the colours gold and bluish green being in marked contrast, and very beautiful.

ε Delphini was known to the Chinese by the unattractive title "the Rotten Melon," certainly a singular and inappropriate name for any star.

Delphinus, in spite of the fact that it is one of the smallest constellations, is one of the best known, and its popular title, "Job's Coffin," has a wider vogue than the great majority of the constellation names.

[1] The star and dolphin combined appear on a coin of the ancient people of Apulia who inhabited Italy before the time of the Romans.

Draco
The Dragon

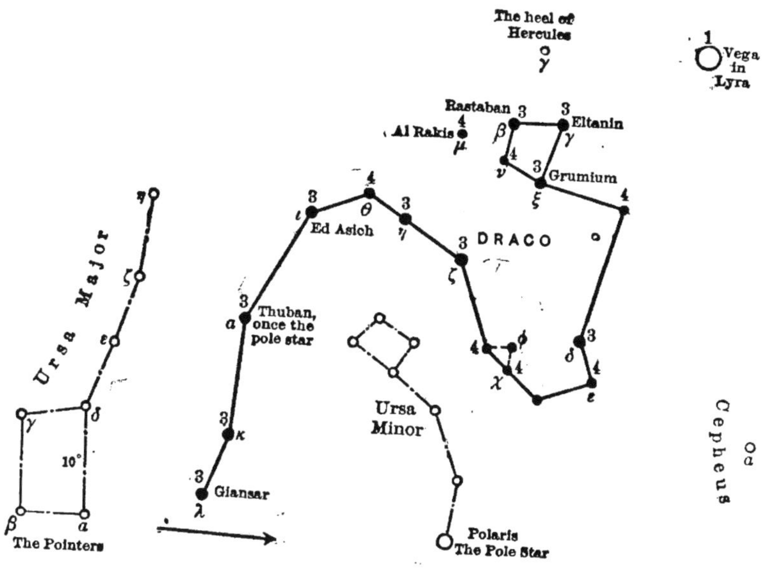
The heel of Hercules
Vega in Lyra
Rastaban
Eltanin
Al Rakis
Grumium
Ed Asich
DRACO
Ursa Major
Thuban, once the pole star
Ursa Minor
Cepheus
10°
Giansar
The Pointers
Polaris The Pole Star

DRACO

DRACO
THE DRAGON

Here the vast Dragon twines
Between the Bears and like a river winds.
WARTON'S *Virgil.*

THERE is no doubt that Draco dates from the earliest times, as Eudoxus and Aratos both mention this constellation, and many believe that it represents the crooked serpent of Job xxvi., 13.

Of all creatures the serpent is historically the most interesting. It is referred to in myth and legend more often than any other, and connected as it is with the very story of Eden, it is linked with the earliest history of man as no other creature is. That it should have found a place amid the constellations is a matter of course, but it seems strange that a creature abhorred instinctively by men of all ages should have been so highly esteemed by the inventors of the constellations as to have been placed by them at the very throne of the heavens.

Burritt says: "Whoever attends to the situation of Draco, surrounding as it does the pole of the ecliptic, will perceive that its tortuous windings are symbolical of the oblique course of the stars. Draco also winds around the pole of the world as if to indicate in the symbolical language of Egyptian astronomy the motion of the pole of the equator around the pole of the ecliptic produced by the precession of the heavens."

In all probability, the twelve signs of the zodiac were the first star groups to be mapped out. The northern stars next claimed attention, and according to Irving,[1] out of these stars was traced a great winged dragon which was supposed to guard the pole of the heavens. In after years the precession of the equinoxes forced the creature off the pole, and he was said to have been overcome by the stalwart Michael, and thrown into a bottomless pit. One writer says: "His tail drew the third part of the stars of heaven, and did cast them to the earth."

The Egyptian hieroglyph for the heavens was a serpent whose scales denoted the stars. When astronomy first began to be cultivated in Chaldea, Draco was the polar constellation.

"It is not known just why this constellation got its fearsome symbol," says Maunder; "the dragon or snake was amongst all ancient nations used to symbolise the powers of evil or darkness or of chaos, but this gives us no explanation why a constellation far from being the least beautiful and conspicuous has been chosen to convey the idea of darkness, still less why such a symbol should have been planted at the very crown of the celestial sphere."

To the early Chaldeans the body of Draco was probably much larger than is now conceded. It surrounded both Bears, and extended downward and in front of Ursa Major.

The Babylonians regarded Draco as a monster personifying primeval chaos—a monster that was finally overcome by a great wind, which was driven with such force into his open jaws that it split him in two.

Brown claims that this constellation is Phœnician in origin, and represents primarily the old serpent, the tempter of Eve in the Garden. Dr. Seiss takes this view of it also.

Mythological accounts of Draco vary considerably. By some this serpent is the guardian of the stars (the golden apples) which hang from the Pole tree in the Garden of Darkness, or Garden in the West, the Garden of Hesperides, near Mount Atlas in Africa.

[1] How to know the heavens by Edward Irving.

Photo by Anderson

Minerva
Vatican Museum, Rome

In north temperate latitudes this constellation never sets, and the Greeks therefore saw in it an emblem of eternal vigilance, symbolised by this Dragon of the Garden, guarding the precious fruit. Juno, it is said, presented these golden apples to Jupiter on the day of their nuptials, and rewarded Draco for his faithful services by placing him among the stars. The legend relates that the serpent was slain by Hercules, and in the old maps Hercules is represented as crushing the head of the Dragon under his foot.

Others claim that this was the snake snatched by Minerva from the giants, and whirled to the sky before it had a chance to uncoil, and that thus twisted it sleeps to-day in the heavens around the axes of the world.

According to another story this is the dragon killed by Cadmus, who was ordered by his father to go in quest of his sister Europa, whom Jupiter had carried away, and never to return to Phœnicia without her. Cadmus having slain the dragon, sowed its teeth and reaped a crop of armed men, who presently engaged in mortal combat from which five only survived. These assisted Cadmus to build the city of Bœotia.

There is an allusion to the Dragon in connection with the familiar story of Phaëton and his desperate adventure with the steeds of day: "When Phaëton rode the chariot of the sun, the horses rushed headlong. Then for the first time the Great and Little Bears were scorched with heat and would fain if it were possible have plunged into the water, and the serpent (Draco) which coils around the North Pole, torpid and harmless, grew warm, and with warmth felt his rage revive."

In Egypt the Dragon was called "Typhon." Plutarch tells us that the hippopotamus, or its variant the crocodile, was certainly one of the forms of Typhon. On the planisphere of Denderah, and the walls of the Ramesseum at Thebes, these animals appear in the circumpolar region, and show clearly that they owe their position to the old myth that the rising sun destroys the circumpolar stars; and Horus, the great god, the light of the heavens, is represented as destroying the hippopotamus or crocodile or Draco. The same idea has come down to us in the well-known myth of St. George and the Dragon.

In Greece Draco was called "Pytho"; in India "Kalli Nagu," meaning the banishment of Vishnu. In Anglo-Saxon chronicles he is referred to as "the fire drake," "the denier of God," "the unsleeping poison-fanged monster," and "the terrible enemy of man full of subtility and power." "The Dragon wing of night overspreads the earth" is an expression which shows the effect of imagination when aroused by the story of such monsters.

There seems to have been a special effort on the part of the originators of the constellations, at the outset almost, to symbolise by a star group the presence of the Evil One, ever watchful, ever vigilant, gazing down upon mortals from the high heavens, as a perpetual menace to evil-doers and a continual reminder of original sin.

The constellations Draco and Hercules are closely associated in ancient mythology, and Hercules is always represented as trampling the Dragon underfoot. These two constellations are in turn connected with Ophiuchus and Serpens, the figure of another giant overcoming a serpent, while he crushes the Scorpion under his feet. On the old maps the figures of these two famous giants appear head to head.

These similar and striking groups, placed so close together in the sky, show clearly that there was a deliberate intention on the part of the inventors of the constellations to emphasise the great fact of a struggle between mankind and serpentkind. There seems here an evident reference to God's interview with the serpent in the Garden of Eden. "I will put enmity between thee and the woman, and between thy seed and her seed. It shall bruise thy head and thou shalt bruise his heel."

There can be little doubt that in these star groups we

have evidence of the very earliest attempts of man to engrave a record of history and tradition for all humanity to read, and that in the history of these constellations lies the key to many of the mooted religious questions of the day.

The Arabs knew Draco as "Al-Tinnīn," and "Al-Thubān," and had names for all the brighter stars in the constellation, many of which represented to their imagination goats or camels. The Egyptians, it is said, also called Draco "Tanen" at one time. This name is still retained by the star γ Draconis.

The dragon was the national emblem of China, but, strange to say, the Dragon of the Chinese zodiac was among the stars composing the constellation Libra. According to Edkins, "the Palace of the Heavenly Emperor" is bounded by the oval formed by the fifteen stars in Draco, amongst which is the star "Tai-yi," the ancient Pole Star, twenty-two degrees from the present Pole.

Schiller thought Draco represented the Innocents, other early Christians saw here the Dragon Infernal.

There are several stars in this constellation that are noteworthy. α Draconis, called by the Arabs "Thubān" and "Al-Tinnīn," was in the year 2790 B.C., or thereabouts, the Pole Star, and the whole constellation then swung around it as on a pivot. Hence it was known in China as "the Right Hand Pivot."

The change in the position of the Pole Star is of such interest, that a slight reference to the reason for it is worth noting. The north celestial Pole is slowly moving in a small circle whose centre is the north Pole of the ecliptic, that great circle which the sun, moon, and planets traverse. This motion causes what is known as "the Precession of the Equinoxes," by which they travel slowly westward, completing an entire revolution in 25,900 years. This motion also causes in time a change in the Pole Star, so that even as Thuban once was, and Polaris now is, so Vega in 13,000 years will mark the Pole.

The importance of the star Alpha Draconis to the ancients is evidenced by the many titles bestowed on it, a few of which are here given: "Judge of Heaven," "High Horned One," "Proclaimer of Light," "High One of the Enclosure of Life," "The Favourable Judge," "Life of Heaven," "the Prosperous Judge," "Crown of Heaven." Seamen regarded it as the Dragon's tail, and Al-Tizini called it "the Male Hyena."

The Great Pyramid was oriented to Thuban, the light from which shone down its central passage in the year 2170 B.C.

According to Maunder, Thuban marked the Pole at the time the constellations were mapped out, a prominence it must have held for over two thousand years. At that time this star must have seemed to all ordinary observation an absolutely fixed centre round which all the other stars revolved, just as Polaris appears to us now. It was however much closer to the true Pole of the heavens than Polaris is at the present time.

Thuban is a 3.6 magnitude star, pale yellow in colour. Herschel claimed that it was formerly a much brighter star than now. With Bayer it was a second magnitude star, and he assigned to it the first letter of the Greek alphabet. It culminates at 9 P.M., June 7th.

γ Draconis, called by the Arabs "Eltanin," the "Dragon's head," vies with Thuban in interest, and has been a notable and much observed star in all ages. It was an object of temple worship in early Egypt, where it was called "Isis." The central passages of the temples of Hathor at Denderah, and that of Thut at Thebes, were oriented to it, the former about 3500 B.C. Long afterwards it served for the orientation of the great temple at Karnak. According to Lockyer, seven other temples were oriented to it, and he considers that the Egyptian goddesses Apet, Mut, Taurt, and Sekhet were the same goddesses under different names, and symbolised the star γ Draconis.[1]

[1] Referring to the matter of orientation Serviss writes: "There is

Photo by Bonfils

Temple of Thebes

Dr. Hooke observed Eltanin with his telescope in the daytime in 1669, but its chief interest lies in the fact that the observation of it led Bradley in 1725 to discover the laws of the aberration of light.

Allen cites the following interesting facts respecting Eltanin: The Bœotian Thebes, the city of the dragon, from the story of its founder Cadmus, shared with its Egyptian namesake the worship of this star in a temple dedicated about 1130 B.C. Eltanin lies almost exactly in the zenith of Greenwich, and hence it has been called "the zenith star." It has been observed at Greenwich for more than two hundred years.

ν Draconis is an interesting double star, separable with a field-glass of high power, the distance of the components being almost exactly one minute of arc.

The remaining stars in the constellation are of no special interest though their many titles are evidence of the great importance of this constellation to the ancients.

---

something magnificent in this thought of the ancient temple-builders to square their work by the stars, and to construct long rows of sphinxes and majestic columns to conduct a ray from the sky to the eye of the god in his dark and hidden chamber, where no impious foot dared follow."

Eridanus
The River Po

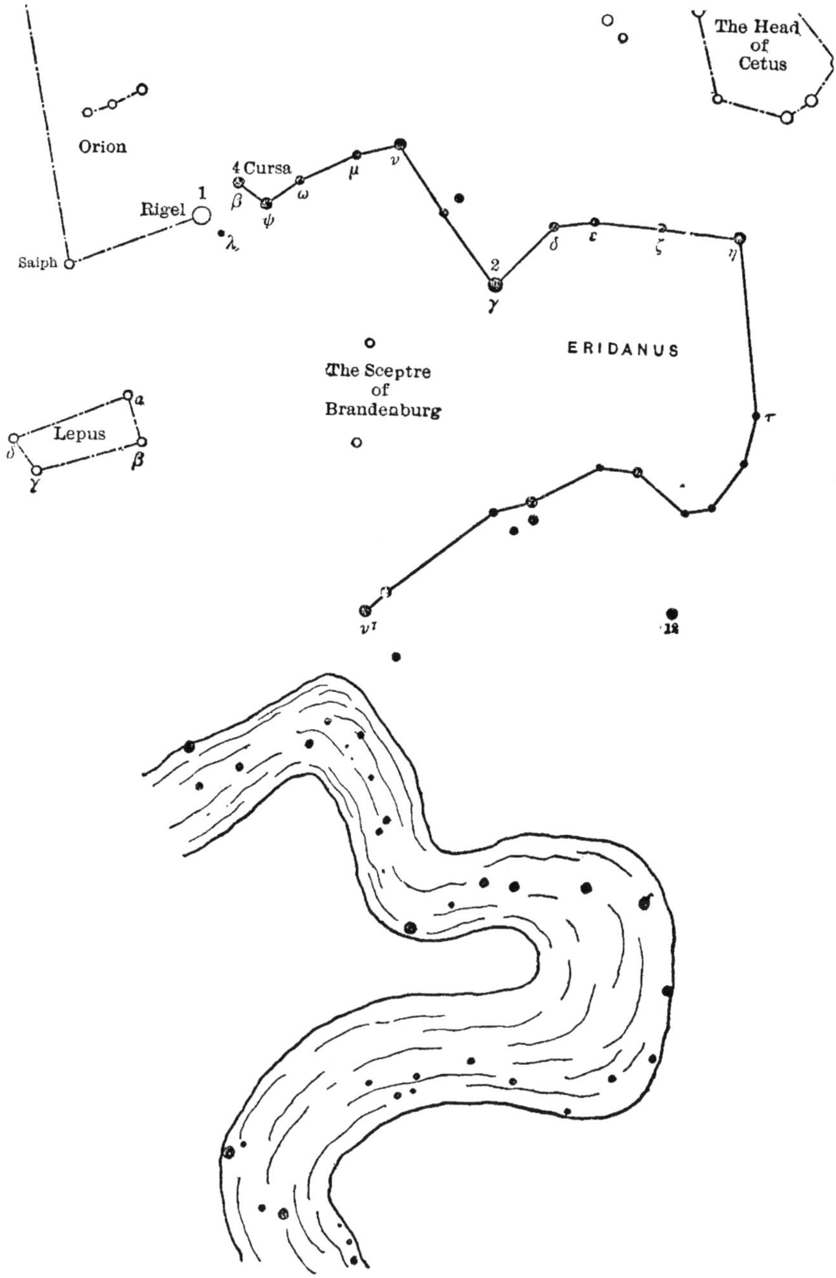
The Head of Cetus
Orion
4 Cursa
β
ψ
ω
μ
ν
1
Rigel
λ
Saiph
2
γ
δ
ε
ζ
η
ERIDANUS
The Sceptre of Brandenburg
Lepus
a
β
γ
δ
τ
ν¹
12

ERIDANUS

ERIDANUS
THE RIVER PO

The scorched waters of Eridanus' tear-swollen flood
Welling beneath the left foot of Orion.

ARATOS.

ACCORDING to Eratosthenes this imaginary river of the stars, winding its devious way across the winter skies, represented the River Nile, and in the Alphonsine Tables it bore the title "Nilus." Brown, however, claims that the Akkadians identified it with the River Euphrates, and that the name "Eridanus" may refer to a Turanian river name meaning "Strong River."

In the Euphratean records there are many allusions to a stellar stream that may refer to this imaginary river, although there is a possibility that the Milky Way is intended, as that always represented a celestial river to the ancients.

Burritt tells us that Eridanus is the name of a celebrated river in Cisalpine Gaul, also called "Padus," the modern "Po." Virgil calls it "the King of Rivers," and the Latin poets have rendered it famous from its connection with the fable of Phaëton, the intrepid youth who endeavoured to drive for a day the chariot of the sun. As the familiar story goes, he was unable to restrain the fiery steeds, and a universal catastrophe was only prevented by a timely thunderbolt from the hand of Jupiter, which hurled Phaëton from heaven into the River Eridanus.

At once from life and from the chariot driven,
The ambitious boy fell thunderstruck from heaven.

His body, consumed with fire, was found by the nymphs of the place, who honoured him with a decent burial. His sisters, the Heliades, mourned his unhappy end, and were changed by Jupiter into poplars, the trees that are found in great abundance in the valley of the Po.

> All the long night their mournful watch they keep,
> And all the day stand round the tomb and weep.
>
> Ovid.

It is said that the tears of the Heliades were turned to amber. Apollonius represented the Argonauts as passing along the banks of the River Eridanus in their voyage from Ister to the Rhone, and as hearing the lament of the Heliades, and seeing their amber tears.

Amber was imported into Greece from the northern shores of the Adriatic, and it was naturally identified with the River Po, the great river of northern Italy.

It is also related that when Hercules went on his quest of the golden apples of Hesperides, he came to the River Eridanus, and enquired his way of the nymphs dwelling near-by.

Burritt thinks that the fable of Phaëton alludes to some extraordinary period of drought and heat which was experienced in a very remote time, and of which only this confused tradition has descended to later times.

The constellation Eridanus is so extended that it has been divided, for the sake of convenience, into a northern and southern stream. The former has its source near the first magnitude star Rigel, in the foot of Orion, and hence Eridanus has been sometimes called "the River of Orion."

Maunder considers that if we regard Eridanus as representing the Flood, and the sacrifice of Andromeda a means to cause its abatement, then Eridanus would stand for the Great Deep of the Primeval Chaos, of which the sea monsters typified the indwelling principle.

The Arab name for this constellation was "Al-Nahr," meaning the River, and the Arabs also imagined that the

Phaëthon Driving the Chariot of Apollo
Painting by Max Klepper

stars in this group represented ostriches, young and old, eggs, and egg-shells.

"The River Jordan," "the Red Sea," and "the River of the Judge" are other names for this famous stream, and it has been identified with Homer's stream flowing around the earth, and sometimes bore the titles "Oceanus," and "the River of Ocean."

β Eridani, called by the Arabs "Cursa," signifying a footstool, is the principal star in the constellation seen in these latitudes. It owes its name to its position close to the foot of the Giant Hunter Orion. The Chinese called this star "the Golden Well," and it was regarded by the Arabs as an ostrich nest, a number of which are scattered through the constellation.

The star γ Eridani was called by the Arabs "Zamack," meaning the "bright star of the boat." This would seem to infer that some sort of a craft was supposed to traverse the stream, and might be an allusion to the Ark, if the stream was originally intended to represent the Flood, as many authorities think.

We have, then, in the constellation Eridanus the diverse representations of a river on which there is a boat, and a gathering place for ostriches, with their nests, eggs, and egg-shells in evidence. This confusion of stellar imagery is one of the features of constellational study, and is at first blush difficult to account for. It seems reasonable to suppose, however, that the art of stellar representation was not confined to any one country, or a particular tribe, and as the nomadic herdsmen travelled from place to place, they left in each a smattering of the star lore that was as much a part of their lives as a knowledge of flocks and herds. As they would in all likelihood see in the star groups whatever their individual fancy dictated, a diversity of representations would naturally follow, which accounts in part for the confusion of figures that a close study of the constellations reveals.

Eridanus is an inconspicuous constellation, but on a

clear night in midwinter, when the moon is absent from the sky, the stream can be traced without difficulty, spreading out like a great horseshoe south and west of the well-known constellation Orion.

Gemini
The Twins

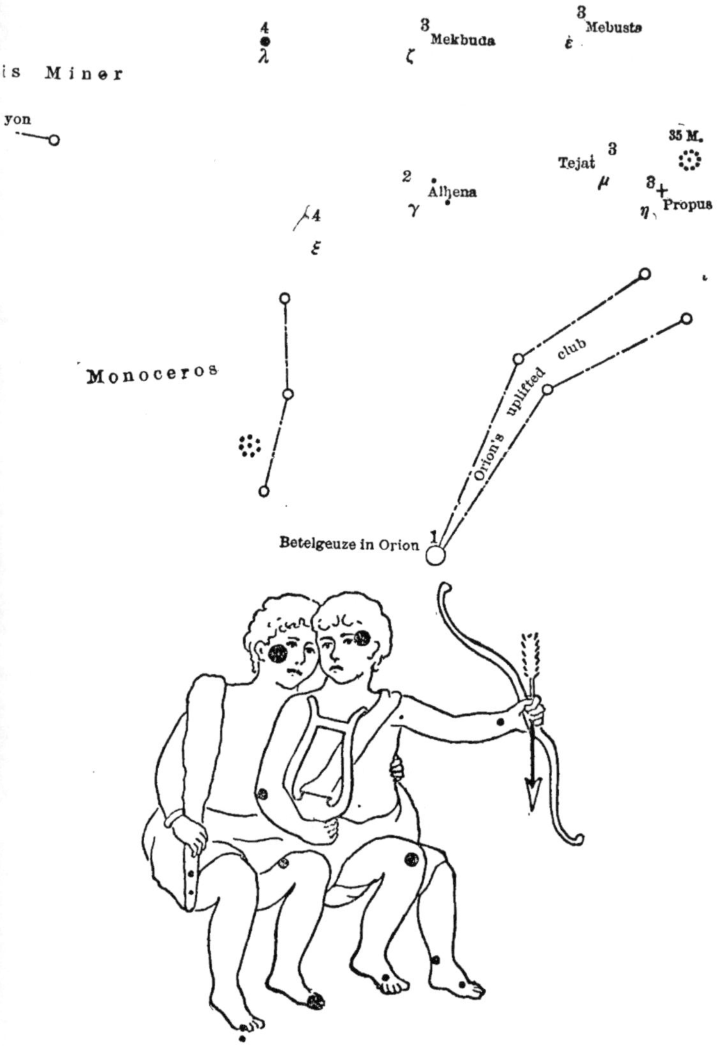

4
λ
3
Mekbuda
ζ
3
Mebusta
ε
is Minor
yon
35 M.
Tejat
3
μ
2
Alhena
γ
3
Propus
η
4
ξ
Monoceros
Orion's uplifted club
Betelgeuze in Orion
1

GEMINI

GEMINI
THE TWINS

> And starry Gemini hang like growing crowns,
> Over Orion's grave low down in the west.
>
> TENNYSON'S *Maud.*

THE ancient Chaldeans, and eastern nations generally, knew nothing of the zodiacal sign we call Gemini, or the Twins, although these stars have always been regarded as twins from remote antiquity.

Instead of twin brothers, however, the ancients imagined these stars represented two Kids. There was a significance in this title quite apart from its relation to the herds that they were daily concerned with.

We see in this region of the sky three ancient and important constellations named after domestic animals that figured prominently in the pastoral life of early times, the Ram, the Bull, and the Kids. Pluche tells us that "in the reproduction of species among the herds familiar to primitive man, the first produced in the vernal season are the lambs, then come the calves, and later the kids, so that it was natural that the ancients who devised the constellations should characterise in this order the three constellations through which the sun passed in the vernal season."

Brown considers that the constellations were designed to perpetuate the stories in which the ancients dramatised their conception of solar and lunar relations. He holds that Gemini is a stellar representation of the great Twin Brethren of the sky, the sun and the moon, who join in building a mysterious city. Although hostile to each other, they work together, and are only seen together by day.

In that interesting book by Maunder, entitled *The Astronomy of the Bible*, we are told that on the Babylonian monuments and boundary stones, the most ancient records extant, there appears a set of symbols repeated over and over again, and always given a position of prominence. It is the so-called "Triad of Stars," a crescent lying on its back and two stars near it.

The significance of this symbol is now clear. Four thousand years before the Christian era, the two stars Castor and Pollux, α and β Geminorum respectively, served as indicators of the first new moon of the year, just as the star Capella did two thousand years later. The "Triad of Stars" then is simply a picture of what men saw year after year in the sunset sky, at the beginning of the first month 6000 years ago. It is the earliest record of an astronomical event that has come down to us.

Plunket says that the early astronomers who mapped out the zodiac, noticing that the equinoctial colure in 6000 B.C. passed the two bright stars Castor and Pollux, chose to represent them as marking the heads of twin figures, which they determined should symbolise the equal day and night of the season over which they presided.

Thousands of years after these two stars had ceased to mark the equinox, they were still associated by the Greeks with the twin heroes Castor and Pollux, brothers, who according to the legend were "possessed of an immortality of existence so divided among them that as one dies the other revives."

The learned Dr. Barrett has pointed out that this furnishes a complete description of day and night, a simile that is especially interesting if we attribute the first symbolising of day and night by these stars to the work of astronomers at a date when the days and nights these stars symbolised were exactly of equal length, and when therefore the equally bright stars and equal alternations of light and darkness might both be fitly symbolised as twins.

The Latin title "Gemini" by which we know the constel-

Photo by Anderson

Temple of Castor and Pollux at Rome

lation dates only from classical times. Burritt gives the following mythological history of the constellation: "Castor and Pollux were twin brothers, sons of Jupiter, by Leda, the wife of Tyndarus, King of Sparta. They were educated at Pallena, and afterwards embarked with Jason in the celebrated contest for the golden fleece at Colchis, on which occasion they behaved with unparalleled courage.

Pollux distinguished himself by his achievements in arms and personal prowess, and was a famous pugilist. Castor was superior in equestrian exercises, and the management of horses. The Twins are represented in the temples of Greece, on white horses, armed with spears, riding side by side."

Among the ancients, and particularly among the Romans, there prevailed a superstition that Castor and Pollux often appeared at the head of their armies, and led on their troops to battle and victory.

> The gods who live for ever
> Have fought for Rome to-day,
> These be the great Twin Brethren
> To whom the Dorians pray.
> Back comes the chief in triumph
> Who, in the hour of fight,
> Hath seen the great Twin Brethren
> In harness on his right.
>
> Macaulay.

Castor and Pollux were a common object of adjuration among the Romans, and the slang of the present day, "By Jiminy," is a survival of the old Roman oath. As guardians of Rome the Twins were inscribed on the Roman silver coins. The "Pence" of the good Samaritan bore their figures, where they were represented as two horsemen. They also appear on coin types of as early date as from 431 to 370 B.C.

Virgil thus writes concerning these illustrious Twins:

> Castor and Pollux first in martial force,
> One bold on foot, and one renowned for horse.

And Martial in like vein:

> Castor alert to tame the foaming steed
> And Pollux strong to deal the manly deed.

After returning from Colchis the brothers waged a successful war against the pirates who infested the Hellespont, from which circumstance they have ever since been regarded as "the sailor's stars," and the friends and protectors of navigation. It is related that Neptune had rewarded their brotherly love by giving them power over wind and wave, that they might assist the shipwrecked.

> Safe comes the ship to Haven
> Through billows and through gales,
> If once the great Twin Brethren
> Set shining on the sails.
>
> Macaulay.

In the Argonautic expedition, during a violent storm, it is said two flames of fire, "St. Elmo," or "St. Helen's light," were seen to play around the heads of Castor and Pollux, and immediately the tempest ceased, and the sea was calm. In honour of the Twins, these lights were sometimes known as "Ledean lights," and sailors believed that whenever both fires appeared in the sky, it would be fair weather, but when only one appeared, there would be storms.

In the *Odes* of Horace, Mr. Gladstone's translation, we read:

> So Leda's twins, bright shining, at their beck
> Oft have delivered stricken barks from wreck.

Homer in his Hymn to Castor and Pollux thus alludes to their supposed influence over the sea:

Photo by Brogi

Temple of Castor and Pollux at Girgenti

These are the Powers who earth-born mortals save
And ships, whose flight is swift along the wave
When wintry tempests o'er the savage sea
Are raging, and the sailors tremblingly
Call on the Twins of Jove with prayer and vow,
Gathered in fear upon the lofty prow,
And sacrifice with snow-white lambs, the wind
And the huge billow bursting close behind,
Even then beneath the weltering waters bear
The staggering ship—they suddenly appear,
On yellow wings rushing athwart the sky,
And lull the blasts in mute tranquillity,
And strew the waves on the white ocean's bed,
Fair omen of the voyage; from toil and dread
The sailors rest, rejoicing in the sight,
And plough the quiet sea in safe delight.

Shelley's translation.

The appearance of the Twins, Castor and Pollux, was hailed as the harbinger of fair summer weather, and they were symbolised by the figure of two stars over a ship.

In the Acts of the Apostles we read that St. Paul sailed from Malta to Syracuse in an Alexandrian ship whose sign was Castor and Pollux, and among the Romans it was very common to place the effigies of the Twins in the prows of vessels.

Castor and Pollux became enamoured of the betrothed daughters of Leucippus, brother of Tyndarus, and resolved to supplant their rivals. A battle ensued, in which Castor killed Lynceus, and was himself killed by Idas. Pollux thereupon killed Idas, but being himself immortal, and most tenderly attached to his brother, he was unwilling to survive him. He therefore besought Jupiter to restore Castor to life. Jupiter granted his request, and made Castor immortal. Consequently as long as one was upon earth, so long was the other detained in the infernal regions, and they thus alternately lived and died every day. As Homer puts it:

By turns they visit this etherial sky,
And live alternate and alternate die.

This idea bears out the alternation of daylight and darkness, and the analogy between the days and nights of equal duration before mentioned.

Jupiter further rewarded their fraternal attachment by changing them both into a constellation, under the name of "Gemini," the Twins, which it is strangely pretended never appear together, but when one rises the other sets, and so on alternately.

Castor and Pollux were worshipped both by the Greeks and Romans, who sacrificed white lambs upon their altars to them. In the Hebrew zodiac the constellation of the Twins refers to the tribe of Benjamin; and according to Dr. Seiss, the Gemini represented the mystic union of Christ and His redeemed. Schiller regarded the constellation as representing St. James the Elder.

The Egyptians represented the Twins as the two gods, Horus, the Elder, and the Younger, and strangely enough also regarded them as Two Sprouting Plants.

The Gemini have also been called "David and Jonathan," "Adam and Eve," "the Twin Sons of Rebecca," Jacob and Esau. The Eskimos recognise in Castor and Pollux the two door-stones of an igloo, the name for their snow huts. The Arabs regarded these twin stars as two Peacocks, and on the Euphratean star list they appear as "the Great Twins," and "the Heaven and Earth Pair."

Allen tells us that in India they always were prominent as "the Aswins," or "Horsemen," a name also found in other parts of the sky for other Hindu twin deities. A Buddhist zodiac had in their place a woman holding a golden cord. Castor and Pollux were regarded as twins by the Assyrians, Babylonians, and the aborigines of the South Pacific Islands.

In Australia they were called "the Young Men." The South African Bushmen on the contrary called them "the Young Women, the wives of the eland," their great antelope.

There appear on the Peruvian star chart of Salcamayhua two figures that resemble the Gemini, and one of the symbols for the Twins was a Pile of Bricks, referring to the building of the first city.

Astrologically considered, this constellation was most favourably regarded, portending genius, goodness, and liberality. It is of the House of Mercury and its native will be tall and straight, with dark eyes, brown hair, and active ways; in character versatile, contradictory, and unselfish. It governs the arms and shoulders and rules over the south-west parts of England, America, Flanders, and Lombardy. It is the ruling sign for those born between May 20th and June 21st. The flower is the Mayflower or trailing arbutus, and the gem, the beryl.

In the early Chinese solar zodiac this constellation figured as "the Ape," and the Chinese astrologers claimed that if Gemini was invaded by Mars, war and a poor harvest would ensue.

Aristotle has left an interesting record of the occultation at two different times of some of the stars of Gemini, by the planet Jupiter, the earliest observations of this nature of which we have knowledge, and made probably about the middle of the fourth century B.C.

No reference to Castor and Pollux would be complete without quoting Mrs. Martin's tribute to them in *The Friendly Stars:*

"The constellation Gemini is the third spring sign of the zodiac, and it is easy to see how the mere beauty of its chief stars, Castor and Pollux, may have fastened upon it the reputation of responsibility for the beautiful weather that comes early in June. At this season of the year position, atmosphere, and surroundings all combine to enhance the beauty and accentuate the individuality of these two beautiful stars. . . . In a comparatively starless environment the twin stars, beloved of sailors, dominate the western sky and shine side by side like two eyes benignly set to keep a protecting watch upon the world. It is not the sailor alone whose fancy is pleased with the kindly vigil they seem to keep. A landsman, too, may have pleasanter dreams if he will but peep through the western window and exchange friendly glances with them before settling down for the night."

Pollux is now the brighter of the two stars, although three hundred years ago, Castor was probably the lucida of the constellation.

Castor is a beautiful star, in fact Sir John Herschel called it the largest and finest of all the double stars in our hemisphere. In a three-inch telescope with a power of ninety diameters, these twin suns present a charming appearance. This is a binary system, with a period somewhere between 250 and 1000 years. According to Allen this star is approaching the earth at the rate of 18.5 miles a second, while Pollux is receding from us at the rate of one mile each second.

Pollux is fifty-four light years distant, and is one of the stars much used in navigation in taking lunar observations. In astrology it was a fortunate star, portending eminence and renown.

The Twin Stars are 4½° apart and this distance was known to the Arabs as "the Ell," a measure of length. In reality Pollux is two hundred trillions of miles farther from us than Castor. There is only eleven minutes difference in the time of culmination of Castor and Pollux, so they may both be regarded as on the meridian at 9 P.M., Feb. 24th.

γ and ξ Geminorum represented to the Arabs the brand made by a hot iron on the neck of a camel, also the star which shines with a sharp light.

δ Geminorum is a double star, known to the Arabs as "Wasat," meaning the "Middle," *i. e.*, of the constellation, says Allen. The Chinese called it "Ta Tsun," the "Great Wine Jar." Just north of this star is the radiant point of the meteors known as "the Geminids," visible early in October.

ε and ζ Germinorum bear respectively the Arab names "Mebsuta" and "Mekbuda."

According to Allen η Geminorum bears the name of "Propus." On Burritt's map this star name is given to a fifth magnitude star a few degrees south and west of η. Eta is noted as marking the locality where Sir William Herschel discovered the planet Uranus, on the 13th of March, 1781. He thought at first that it was a comet, and reported it as such. Maskelyne, however, suspected its planetary nature, and the succeeding year it was announced as a new planet by Lexell and La Place.

"Continental astronomers designated the planet as 'Herschel' and we find this title in text-books as late as fifty years ago. Bode suggested the present title Uranus, to conform to the mythological nomenclature of the other planets, and because the name of the oldest god was especially applicable to the oldest, the most distant body then known to our system."

R. H. ALLEN, *Star Names and Their Meanings.*

There is a star of the fifth magnitude, just west of μ Geminorum, which is noteworthy as marking the location of the summer solstice, in the tropic of Cancer, the place occupied by the sun on the longest day of the year, and is moreover the dividing limit between the torrid and north temperate zones.

Gemini contains a beautiful star cluster in M 35. La Place thus describes this magnificent object: "A marvellously striking object. No one can see it for the first time without an exclamation. . . . The field is perfectly full of brilliant stars, unusually equal in magnitude and distribution over the whole area. Nothing but a sight of the object can convey an adequate idea of its exquisite beauty."

14

The Triad of Stars
From a Babylonian Boundary Stone
Approximate date 1200 B.C.

Photo by Brogi

Farnese Hercules
National Museum, Naples

Hercules

The Kneeler

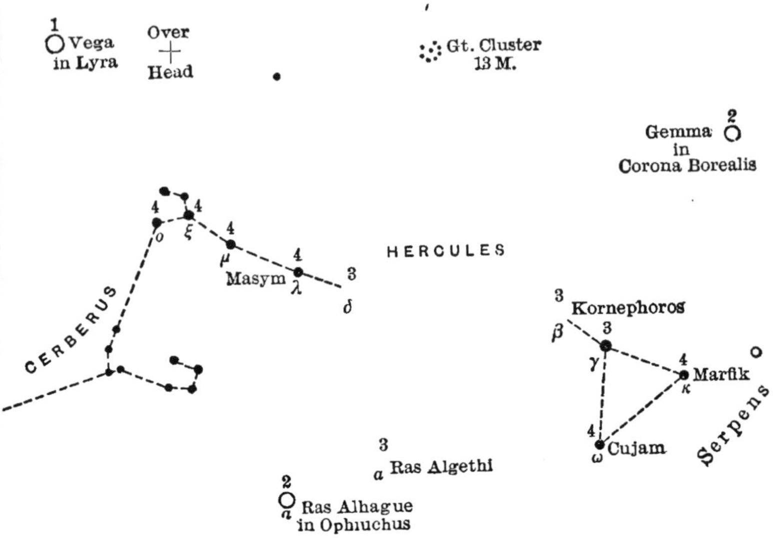
1
Vega
in Lyra
Over
Head
Gt. Cluster
13 M.
2
Gemma
in
Corona Borealis
4
4
o
ξ
4
μ
HERCULES
4
Masym
λ
3
δ
CERBERUS
3
Kornephoros
β
3
γ
4
Marfik
κ
Serpens
4
Cujam
ω
3
a Ras Algethi
2
a Ras Alhague
in Ophiuchus

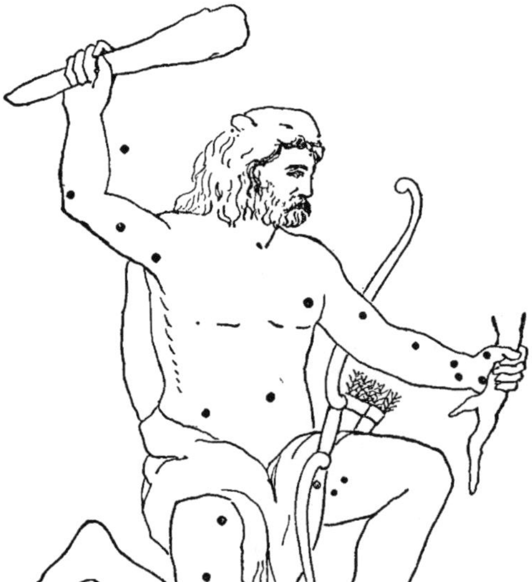

HERCULES

HERCULES
THE KNEELER

> An Image none knows certainly to name
> Nor what he labours for.
>
> ARATOS.

IF a variety of titles are indicative of the antiquity of a constellation, and its importance in the minds of the ancients, the constellation Hercules may well be considered one of the oldest and most prominent of the early star groups.

The origin of this constellation is shrouded in mystery. It was not known to Greek astronomers by the name "Hercules," but as "Engonasi" or "Engonasin," meaning the "Kneeling One." They also knew it as "the Phantom," and "the Man upon his Knees." Aratos refers to the figure as "the inexplicable Image." Manilius also alludes to the mystery attached to the title of this star group. Creech thus translates the passage:

> Conscious of his shame
> A constellation kneels without a name.

As Hercules is usually represented,

> . . . his right foot
> Is planted on the twisting serpent's [Draco's] head.

In his right hand he brandishes a club, and in his left he holds a branch, in which serpents are entangled. Over his shoulders is thrown a lion's skin.

Allen tells us that some modern students of Euphratean mythology associate Hercules and Draco with the Sun-god Nimrod, and the dragon Tiamat slain by him. This tradition probably served as the foundation of the classical myth concerning Hercules and the Lernæan Hydra..

Burritt tells us that this constellation was intended to immortalise the name of Hercules the Theban, the son of Jupiter and Alcmena. By the poets Hercules was often referred to as "Alcides," possibly derived from Alcæus, the name of his grandfather.

Even in his infancy, Hercules displayed great courage and strength, for it is related that he rose in his cradle and strangled the serpents sent by Juno to destroy him. He was educated by the centaur Chiron, and when eighteen years of age commenced a career that was destined to immortalise him. Hercules was subjected to the will of Eurystheus, and at his behest performed the wonderful feats of strength and agility that have been universally known as "the twelve labours of Hercules." In addition to the performance of these arduous duties he found time to accompany the Argonauts to Colchis, assisted the gods in their war against the giants, conquered Laomedon, and pillaged Troy. Unfortunately the hero was subject to fits of insanity, and when these seized him performed many rash deeds. One of these was his attempt to carry off the sacred tripod from Apollo's temple at Delphi. As a punishment for his misdeeds he was sold as a slave to Queen Omphale of Lydia, and it is said was often observed spinning with the Queen's maidens in the women's hall; subsequently he married the Queen.

He re-established his friend Tyndarus on the throne of Sparta, and for his second wife married Dejanira, a sister of Meleager, and took up his abode in the court of Ceyx, King of Trachina. As Hercules was setting out on some journey his wife unwittingly presented him with a cloak which she had received from the centaur Nessus, whom Hercules had slain for insulting Dejanira. This garment proved to be poisoned. Hercules was infected, and feeling his end approaching, resolved to die a death worthy of a famous hero. He erected a great funeral pyre on Mt. Œta.

Photo by Brogi

The Infant Hercules Strangling the Serpents at Pompeii

Calmly taking his place upon it, the torch was applied, and he suffered the death by fire resignedly. After his body was consumed, the ancient poets say, he was carried up to heaven in a chariot drawn by four horses:

> . . . 'Almighty Jove
> In his swift car his honour'd offspring drove;
> High o'er the hollow clouds the coursers fly,
> And lodge the hero in the starry sky.
>
> Ovid.

The admiring gods gave him Hebe, the cup-bearer of the immortals, as his wife.

Hercules, because of his great physical prowess and his success in accomplishing well-nigh impossible feats, was one of the most popular of the figures of mythology. The Fabian gens of Rome, a race of men superior in physical and intellectual attainments, claimed that they were descended from this paragon of fearlessness.

The twelve labours of Hercules are supposed to have an astronomical significance, and to refer to the sun's passage through the zodiacal signs. "Beginning with the summer solstice a series of coincidences will be noted which makes impressive this ancient belief. For example the first sign through which the sun passes is Leo, and Hercules' first labour was the slaying of the Nemean lion. In the second month," says Anthon, "the sun enters the sign Virgo, when the constellation of the Hydra sets, and in his second labour Hercules destroyed the Lernæan Hydra. In the third month the sun enters the sign Libra, when the constellation of the Centaur rises, and in his third labour Hercules encountered and slew the centaur. These comparisons are traceable throughout the year and add distinct testimony to the ingenuity of the ancients." For a more detailed account of this matter the reader is referred to Anthon's Classical Dictionary.

There is certainly a significance in the location of this figure of a giant trampling on a serpent, for he is placed head to head with the giant Ophiuchus, who is represented as holding a writhing serpent in his grasp. Hercules has been thought to represent the first Adam, beguiled by the serpent, and condemned to a life of toil, while Ophiuchus is supposed to be the second Adam, triumphant over the serpent.

The relations of mankind and serpentkind dwelt on in the Bible, and figuring so prominently in the sky figures, seems to indicate the antiquity of these constellations, and shows clearly that there was a deliberate plan carried out in designing them. The constellation Hercules is without doubt linked with the earliest records of the history of man.

In Phœnicia, the constellation Hercules is said to have represented the god Melkarth, and was an object of worship, Melkarth being regarded as a Saviour by the Phœnicians. It has also been identified with Ixion, Prometheus Bound, and Theseus.

Brown holds that the constellation is Euphratean in origin, and was known originally as "Lugal," or "Sarru," the King.

It is significant that we always find Hercules represented as kneeling, an incongruous position for a hero or god engaged in trampling on a great serpent. There appears no satisfactory explanation for this attitude. Blake says that there is a story related by Æschylus about the stones in the Champ des Cailloux, between Marseilles and the embouchure of the Rhone, to the effect that Hercules being amongst the Ligurians, found it necessary to fight with them, and looked about in vain for some missiles to hurl at his foes. Jupiter, touched by the danger of his son, sent a rain of round stones with which Hercules repulsed his enemies. The Engonasis is thus considered by some to represent the giant bending down to pick up the missiles.

In the modern representations of the figure, Hercules swings a club in one hand, and holds fast a branch, or the three-headed dog Cerberus, in the other, so that there does not seem much reason or opportunity for him to pick up

Photo by Neurdein

stones. Posidonius sagely remarks that it was a pity Jupiter did not rain the stones on the Ligurians in the first place, and save Hercules the trouble of picking them up.

Bayer represents Hercules as holding in addition to his club an apple branch, possibly to indicate his connection with the myth of the Golden Apples of the Garden of Hesperides. For his eleventh labour he was ordered to procure them.

Those who claim that Hercules represents Adam certainly have much to substantiate their theory, for associated with the figure we find a serpent, a garden, and the apple.

"In latitude 40° north, 4667 B.C.," says Plunket, "Hercules culminated gloriously on the northern meridian at midnight of the spring equinox. Never since that date has he held so commanding a position in the sky." At the present time and in our latitude Hercules will ever rise reversed, and through the summer and autumn months his kneeling figure is always to be seen hanging downwards in the sky in anything but a dignified or commanding attitude. We may readily suppose that those who beheld this grand and conquering figure considered that it typified the ever increasing triumph at that season of the year of the power of light over darkness.

The Greek name "Herakles," for which there appears no Aryan derivation, is a rendering of the Phœnician "Harekhal" (the traveller), the Latin Hercules. Herakles was represented on coins of Cyzicus, about 500–450 B.C. It is perhaps the most familiar coin type throughout Hellas.

According to Maunder, the first suggestion that this kneeler was the great national Hellenic deity seems to have been due to Panyasis, the uncle of the great historian Herodotus. In a poem on the subject of the great national hero, in order to do him the greater honour, he sought to identify him with the unnamed wrestler of the constellation. The fact that, despite this effort, the identification had entirely failed of adoption two hundred years later, is as near positive proof as we can get, not merely that it was not known whom the constellation represented, but that it was known that it did not represent Hercules.

The faint stars in this region of the sky, according to the ancients, represented a meadow, where a shepherd pastured his flock, and the long rows of stars in Hercules and Serpens are fences protecting the sheep from the hyenas and jackals that are supposed to be prowling about. The other side of the meadow is protected by the shepherd's two dogs, the stars α Herculis and α Ophiuchi representing the dogs. To the early Christians Hercules represented the Three Wise Men of the East, and more appropriately Samson.

α Herculis, or Ras Algethi, the Arab name, meaning the head of the kneeler, is a beautiful double star, with a fine contrast of colours in its orange red and bluish green stars. Its variability was discovered by Sir Wm. Herschel in 1795. It is one of the most noted of Secchi's third type with banded spectra. The Chinese called it "the Emperor's Throne."

λ Herculis is noteworthy as marking the approximate objective point, according to Sir Wm. Herschel, of our solar system, the so-called Apex of the Sun's Way, whither we are speeding at the rate of from 7½ to 11½ miles a second. More recent observations show that the goal of our system is situated in the vicinity of the star Vega in the constellation Lyra.

Hercules is remarkable for containing a wonderful star cluster, situated between the stars ξ and η. Halley discovered it in 1714, and thought it a nebula, and "Halley's nebula" was its early title. It can only be resolved in large telescopes. Harvard observers have counted as many as 724 stars outside of the nucleus in this wonderful cluster. It is visible to the naked eye on a clear night, and is considered the finest cluster in the northern heavens, although in a small telescope it does not compare in beauty with the beautiful clusters in Perseus, Gemini, and Sagittarius.[1]

[1] Concerning this wonderful object Serviss writes: "You must go to the southern hemisphere to find its match anywhere in the sky.

Herschel estimated that it contained 14,000 stars. In a recent photograph of this cluster 50,000 stars are shown in an area of the sky which would be entirely covered by the full moon.

Photographs of this swarm of suns fail to do it justice because of its density. In order to give some idea of the wonderful structure of such a cluster a photograph of the globular cluster ω Centauri in the Southern Hemisphere is here shown. It is estimated that this cluster contains in the neighborhood of ten thousand stars, and within its confines Professor Bailey discovered 128 variable stars. This remarkable photograph was taken at the Southern Station of the Harvard College Observatory at Arequipa, Peru, and the writer is indebted to Prof. E. C. Pickering, Director of the Harvard College Observatory for the print here reproduced.

The stars π, ε, ξ, η, Herculis form a figure not unlike a keystone, which serves many as a means of identifying the constellation.

It is a ball of suns. Now you need a telescope. You must have one. You must either buy or borrow it, or you must pay a visit to an observatory, for this is a thing that no intelligent human being in these days can afford not to see. Can it be possible that any man can know that fifteen thousand suns are to be seen, burning in a compact globular cluster, and not long to regard them with his own eyes?"

Photo by Anderson

Hercules and Hesperides
Villa Albani, Rome

Hydra
The Water Snake

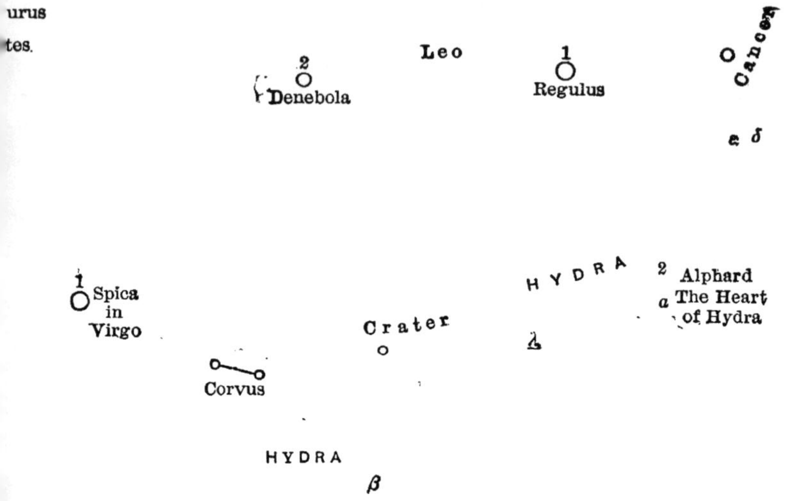
urus
tes.
Leo
2
Denebola
1
Regulus
Cancer
ε δ
HYDRA
2
Alphard
α The Heart of Hydra
1
Spica in Virgo
Crater
Corvus
HYDRA
β

HYDRA

Photo by Prof. Bailey, Harvard College Observatory
Star Cluster in Centauri

HYDRA
THE WATER SNAKE

But lo: afar another constellation,
They call it Hydra like a living creature.
'T is long drawn out. His head moves on below
The midst of the Crab; his length below the Lion,
His tail hangs o'er the Centaur's self.

FROTHINGHAM'S *Aratos.*

BURRITT'S chart represents Hydra as bearing on his coils, as he draws himself along the sky, a Cup, and two birds, the Crow and the Owl, the latter being a recently added asterism. His fierce head and extended fangs seem dangerously close to the Lesser Dog, while the Crab, just above him, lies in wait to seize him in its vice-like claws.

Hydra is supposed to be the snake shown on a uranographic stone from the Euphrates of 1200 B.C., "identified with the source of the fountains of the great deep," and one of the several sky symbols of the great dragon Tiamat. On one of the Euphratean boundary stones, those most ancient records of early times, the figures of the Water Snake and the Scorpion appear side by side.

Brown claims that Hydra is a variant reduplication of Cetus, the storm and ocean monster attacked by the sun-god, and in an ancient Akkadian hymn it is referred to as bearing the yoke on its seven heads.

The ancients perceived an analogy between a quick flowing river, and the swift gliding of a huge glistening serpent, and so, as Maunder says, we arrive at the idea of the River of the Snake, which develops into an ocean stream. The Egyptians at one time regarded this constellation as the celestial counterpart of the River Nile. It has also been identified with the Argonautic constellations.

In Greek mythology Hydra was the Lernæan monster, a great water snake, destroyed by the famous Hercules as the second labour imposed on him. It is related that this fierce serpent lived in a swamp near the well of Amymone, and was wont to ravage the country of Argos. The monster had one hundred heads according to Diodorus, fifty according to Simonides, and nine according to the more commonly received opinion of Apollodorus, Hyginus, and others. The head in the centre was said to be immortal. As fast as Hercules struck off one of the monster's heads with his club two new ones appeared in its place, and the task of slaying the monster appeared hopeless. At this juncture, the legend relates, Iolaus, the faithful nephew of Hercules, came to his assistance, and suggested burning off the heads of the serpent. This they successfully accomplished, and the ninth head which was immortal they buried under a rock. Hercules then dipped his arrows in the Hydra's blood, which ever after rendered mortal the wounds they inflicted.

Juno, jealous of the success of Hercules, sent a sea crab to bite his foot while he was engaged in slaying the Hydra, but the giant easily disposed of the crustacean, much to Juno's disgust.

This myth connects Hydra with the Crab, a relation which, owing to their proximity in the sky, would seem to call for an explanation.

Burritt claims that this fable of the many-headed Hydra may be understood to mean nothing more than that the marshes of Lerna were infested with a multitude of serpents, which seemed to multiply as fast as they were destroyed.

Among the constellations we find the figures of three serpents. At the present time their position in the heavens does not appear especially significant, but in order to understand in a measure the history of the constellations, we must

Photo by Anderson

Hercules and the Hydra
Uffizi Gallery at Florence

endeavour to see the heavens as viewed by the ancients who designed them.

By reason of the Precession of the Equinoxes, that slow change in the position of the heavens, which is constantly going on, the constellations bore different relations to the important points in the heavens than they do now. A Precessional globe enables us to see the star groups as they appeared at any period, and taking the date Maunder suggests, 2700 B.C., as the approximate time when the constellations were designed, and 40 degrees north latitude as the probable abode of those who planned them, we find in the grouping of the serpentine constellational figures several significant facts. The far-extended Hydra crawls along his full length, "going on his belly." The Serpent clasped tightly in the hands of the giant Ophiuchus writhes upward in his struggle to escape, while the Dragon, the great serpent of the north, twines around the crown of the sky, as if guarding the Pole.

Again Hydra lay at this time along the equator, taking seven hours out of the twenty-four to cross the meridian. The Serpent marked the intersection of the equator with one of the principal meridians of the sky, while the Dragon of the north linked the north pole of the celestial equator to the north pole of the ecliptic.

These facts seem significant, and tend to show that there was a definite plan in the minds of those who designed these star groups.

Further, Draco was supposed to represent the oblique course of the stars, while Hydra, the great southern serpent, symbolised the moon's course.

Hydra has been identified with the Flood, the River Jordan, and Plunket claims that it represents the demon Vrita, of the Rig-Veda, conquered by India. Between the first magnitude stars Procyon and Regulus, and between the ecliptic and equator, there is a group of stars in Hydra marking the head of the creature, a striking and conspicuous group, forming a rhomboidal figure. These stars were called by the Chinese "the Willow Branch," or "Circular Garland," which rules over planets, and forms the beak of the "Red Bird." It was worshipped at festivals of the summer solstice as an emblem of immortality.

Here, too, Allen tells us was the seventh Hindu lunar station, known as the "Embracer," which was figured as a wheel. Edkins asserts that this star group was also known as "the Seven Stars." In the Euphratean star list it bears the title of "the Mouth of the Snake Drinks."

According to Dr. Seiss, the Hydra stands for that Old Serpent called the Devil, while to Schiller it represented the River Jordan.

The only star of importance in this constellation is the second magnitude star α Hydræ, a dull red star known to the Arabs as "Alphard," meaning the "solitary one," an appropriate title as there are no other bright stars in this region of the heavens. Tycho Brahe was the first to call it "Cor Hydræ," the heart of Hydra, a familiar name for the star. The Arabs also knew it as "the Backbone of the Serpent," and it was the most prominent star in the great Chinese asterism called "the Red Bird." Alphard culminates at 9 P. M. on March 26th.

Eudosia thus alludes to Hydra and the star Alphard:

> . . . Near the equator rolls
> The sparkling Hydra, proudly eminent
> To drink the Galaxy's refulgent sea;
> Nearly a fourth of the encircling curve
> Which girds the ecliptic, his vast folds involve;
> Yet ten the number of his stars diffused
> O'er the long track of his enormous spires;
> Chief beams his breast, sure of the second rank,
> But emulous to gain the first.

Garrett P. Serviss, who has done so much to popularise astronomy, writes as follows of the stars like Alphard lone and unattended: "There is an attraction about these solitary bright stars that is almost mystical, their very loneliness lending interest to the view, as when one watches some distant snow-clad peak gleaming in the rays of sunset after all the lower mountains have sunk into the blue shadows of coming night."

The star ε Hydræ presents a paradox. It is a double star. The brighter star, of the third magnitude, is sixteen times brighter than its companion, yet the fainter star has six times the mass of the brighter.

Leo

The Lion

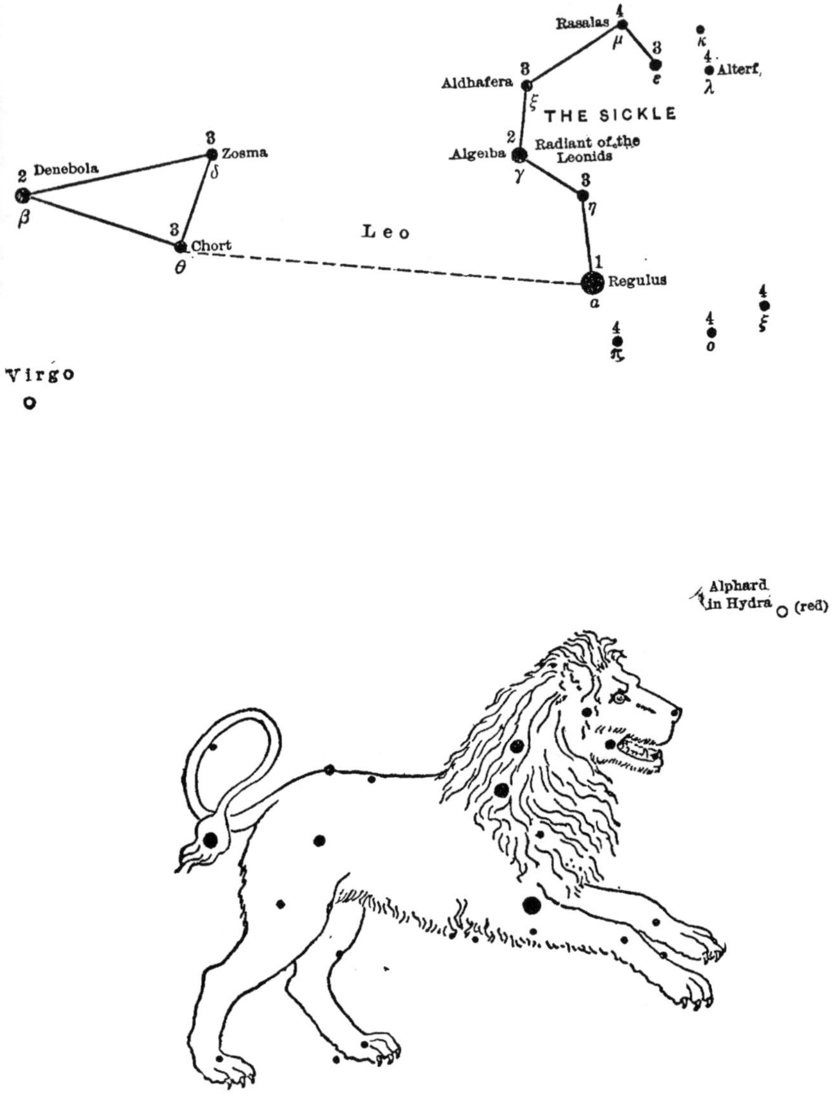
Rasalas
Aldhafera
THE SICKLE
Alterf
Algeiba
Radiant of the Leonids
Zosma
Denebola
Chort
Leo
Regulus
Virgo
Alphard in Hydra
(red)

LEO

LEO
THE LION

The Lion flames. There the sun's course runs hottest.
Empty of grain the arid fields appear
When first the sun into the Lion enters.

ARATOS.

THE figure of Leo, very much as we now have it, appears in all the Indian and Egyptian zodiacs, and of all the zodiacal constellations it is probably the most famous. As many authorities claim, its prominence is beyond question due to the fact that the place of the sun at the summer solstice was in this constellation at the time when the star groups were designed. There was thus a visible connection between the constellation Leo and the return of the sun to the place of power and glory at the apex of the heavenly arch. This obvious relationship is the principal reason why Leo was held in such high esteem and reverence by the ancients.

Owing to the change wrought by the Precession of the Equinoxes, the sun in ancient times entered this constellation about a month earlier than it does now, at a time when the heat of summer was at its maximum. "The sun glows in the Lion," says Seneca, meaning that when the sun enters the sign of Leo at the summer solstice, the highest temperature of the year is experienced. The placing of the fiery and ferocious Lion, the king of beasts, in this part of the sky, symbolised the fact that the sun reigned supreme when it arrived in this constellation.

To escape this season of heat, the lions of the desert sought the valley of the Nile, that river attaining its highest level in the latter part of July, when the sun was in Leo, and this fact is thought by some to account for the name of the constellation.

The figure of a lion's head was usual on the gates which opened the canals irrigating the Nile Valley, and we find, even in modern times, fountains springing from the gaping jaws of graven lions, patterned after ancient fountains, a decorative symbol that was universally employed throughout the Greco-Roman world.

At Athens, Ephesus, Olympia, Agrigentum, and many other places, lion fountains are found, but it is not definitely known where the idea of the design originated. Curtius describes an Assyrian bas-relief from Bairan, showing water streaming from a ring-shaped vessel, and on either side of the fountain, as if on guard, stands the figure of a lion.

The water clock, which was used by the ancients in judicial proceedings, had the form of a lion, and a name which signified "the guardian of the stream," and some think the idea of protection may have been the origin of the association of lions with fountains, and that the custom may have been originated in Asia.

The connection between the sun, king of the heavenly hosts, and the lion, king of beasts, is obvious. Macrobius says: "This beast seems to derive his own nature from that luminary [the sun], being in force and heat as superior to all other animals, as the sun is to the stars." The Lion is always seen with his eyes wide open, and full of fire.

There is little doubt as to the existence of the Lion among the first Babylonian constellations, and throughout antiquity it has held a close relationship with the sun. It was "the Fiery Trigon" of the Arabs. The Egyptians worshipped it because the sun's entrance into the sign coincided with the inundation of the Nile, and some authorities think that the mysterious Sphinx symbolises Leo. The Mexicans also worshipped the Lion, and the chief Druid of Britain was styled "a Lion." The national banner of the ancient Persians bore the device of the sun in Leo, and a lion couchant with the sun rising at his back was sculptured on many of their palaces.

Among the Peruvians, Leo has the form of a puma springing upon his prey, and thus we find the primitive people of the eastern and western world viewing in this region of the heavens a gigantic feline creature.

The lion was the symbol of the tribe of Judah, and the constellation appears in the Hebrew zodiac. It was this tribal symbol of Judah that appeared emblazoned on the shield of Richard I. the Crusader. The association of Leo with Judah arose from the fact that Leo was Judah's natal sign. In the Bible there are frequent allusions to this connection between Leo and the tribe of Judah. Thus we read: "Judah is a Lion's whelp," and again, "The Lion of the Tribe of Judah hath prevailed."

Christians of the Middle Ages called Leo one of "Daniel's Lions," and distinct reference is made to Leo in an inscription on the walls of the Ramesseum at Thebes. To Schiller Leo represented St. Thomas.

According to Greek fable, this Lion represents the formidable animal which infested the forest of Nemæa. It was slain by Hercules, his first labour, and placed by Jupiter among the stars in commemoration of that dreadful conflict. Hercules is generally represented as wearing the lion's skin, and he is said to have reclined on it as he awaited his doom on the funeral pyre. Some aver that Hercules strangled the lion with his hands, but according to another legend he seized the lion by its jaws, and drove his heavy club down the creature's throat.

Maunder points out a curious relationship between four of the zodiacal constellations, one of which is Leo. He says: "The four most important signs of the zodiac are those in which the sun is located on the longest and the shortest days, and on the two days when the days and nights are of equal length. These four signs in the days of Noah were the Bull, the Lion, the Scorpion, and the Water-Pourer.

"The Bull, Lion, and Man are three of the four Cherubic forms frequently referred to in the Bible, and so often an object of worship in early idolatries. The fourth form, the Eagle, is so closely associated with the Scorpion, that it is an evident fact that the guardianship, as it were, of the four quarters of the heavens had been allotted to these four mysterious forms."

The Medes, who dwelt in the vicinity of Babylon early in the fourth millennium B.C., invented an astronomic monogram in which, some claim, there may be clearly read an allusion to these four constellations of the zodiac, which at that date marked the four seasons.

This monogram was used as a standard thousands of years later by the Semitic Assyrians. The principal stars in these four constellations were known to the ancients as "the four Royal Stars."

The Persians had a tradition that four brilliant stars marked the four cardinal points, *i. e.*, the colures, and these Royal Stars were Regulus, in Leo, Aldebaran, in Taurus, Antares, in Scorpius, and Fomalhaut, in the Southern Fish. These four stars were celebrated throughout all Asia. The brilliant star in the Eagle, Altair, has been suggested as the fourth Royal Star instead of Fomalhaut. Thus, as in the vision of Ezekiel, so in the constellation figures, the Lion, the Ox, the Man, and the Eagle stood as the upholders of the firmament, as "the pillars of heaven." They looked down like sentinels upon all creation, and seemed to guard the four quarters of the sky.

Leo is for many reasons significant to Masons. In the four Royal Stars, the four great Elohim, or Decans, gods ruling the signs, were believed to dwell. The four Decans who ruled the four angles of the heavens were the most important and most powerful.

To these four stars divine honours were paid, and sacred images were erected in which the Lion, Eagle, Ox, and Man were variously combined. These figures appear on the

Royal Arch Banner, and the Royal Arch itself is best exemplified by the appearance of the constellations themselves, and Leo, typical of strength, is at the very summit of the Arch at low twelve on Feb. 5th. This is the best time to view the Arch, says Brown,[1] as it then appears in all its beauty in the starry skies.

The symbol of Leo (♌) some think is intended to represent a crouching lion, or its mane or tail; others claim that it outlines the conspicuous figure in the group, the so-called "Sickle of Stars," by which many identify the constellation. The centre of the "Sickle" marks the radiant point of the celebrated Leonid meteor shower, that astonished the world by the brilliant displays of 1833 and 1866, and to which we owe much of our knowledge of meteoric astronomy.[2]

Besides the figure of the Sickle which marks the head of the Lion, there is a rectangular figure which marks his hind quarters. In this place was situated one of the Hindu lunar stations, represented by the figure of a Bed or Couch.

Astrologers distinguished Leo as "the sole house of the sun," and taught that the world was created when the sun was in that sign. They called it "the House of Lions." Those born between July 22d and Aug. 22d are said to be born under the sign Leo and governed by the sun. Such persons are large framed, austere of countenance, with dark eyes and tawny hair, strong voice, and leonine character, resolute and ambitious, but generous and courteous. Leo governs the heart and back and reigns over Italy, France, Bohemia, Sicily, Rome, Bristol, Bath, Taunton, and Philadelphia. It is a masculine sign and fortunate.

[1] *Stellar Theology* by Robt. Brown.

[2] "The 'Sickle' in its entirety," says Serviss, "is an attractive asterism, and hanging so conspicuously in the sky on a spring evening it may be imaginatively regarded as a harbinger of the opening of the season when the thoughts of men are turning to preparations for future harvests."

The morning-glory is the emblematic flower, and the significant stone is the ruby.

Only two Emperors in all history were ruled by Leo. They were Marcus Aurelius and Claudius Gothicus. In a recent article in the *Century Magazine*, entitled "A Discovery concerning Marcus Aurelius," the author bases his alleged discovery of the tomb of this Emperor partly on the fact that on the cover of the sarcophagus there appears the figure of "a lion all alone, a sort of heraldic-appearing lion reclining with paws crossed . . . the lion of the sign of the zodiac." This the author claims was the best evidence that the person buried in the sarcophagus was born under the sign Leo, and from further facts he comes to the conclusion that this tomb was that of the noted Emperor.

The constellation Leo bears little resemblance to the outline of the king of beasts, and some authorities think that the name was originally applied only to the principal star in the constellation, the first magnitude star "Regulus," meaning the "little King" or "Prince." This has also been called "the Star Royal," and the cuneiform inscriptions of the Euphratean Valley refer to it as "the star of the King." The Arabs knew it as "the Kingly Star," and it was one of the four celebrated Royal Stars before alluded to.

Apparently its position and not its lustre has made Regulus famous, for almost all the first magnitude stars exceed it in brilliance.

Regulus has been a famous star in all ages. The ancient belief was that it ruled the affairs of heaven, and with astrologers it was always a fortunate star. According to the best authority, Regulus was not named from the illustrious Roman Consul of that name, as has sometimes been supposed, but was named by Copernicus from the diminutive of the earlier "Rex."

The impression of greatness and power connected with Regulus was universal. This was doubtless due to the fact that it was the brightest star in the principal zodiacal sign.

"Cor Leonis," or "the Heart of the Lion," was another name for this star, and Al-Biruni called it "the Heart of the Royal Lion."

The importance of Regulus in ancient times is well attested by the great variety of names assigned it, titles for the most part signifying power and might. In Babylon it was "Sharru," the King, in India, "Magha," the Mighty, in Sogdiana "Magh," the Great, in Persia, "Miyan," the Centre, among the Turanians "Masu," the Hero, and in Akkadia it was associated with the fifth antediluvian King of the celestial sphere. In Arabia it was known as "Kingly," in Greece βασιλισκός ἀστήρ, the equivalent of Rex, the King Star.

On a Ninevite tablet there is this reference to Regulus: "If the star of the great lion is gloomy, the heart of the people will not rejoice."

Regulus is one of the so-called "Lunar Stars," and is consequently much used in navigation. On the 20th of August Regulus almost marks the position of the sun. It has a spectrum of the Sirian type, and is approaching the earth it is said at the rate of 5.5 miles a second. Some authorities claim that this great sun sends out a thousand times as much light as our sun, and is 160 light years distant from us.

Mrs. Martin, who has endowed the first magnitude stars with an individuality that will ever enhance their beauty, and endear them to all star lovers, regards Regulus as the most neighbourly of stars, as it is visible for eight months in the year. The following reference to Regulus is quoted from Serviss's *Round the Year with the Stars:* "When the 'Royal Star' crosses high on the meridian in the vernal evenings, the imagination is thrown back almost the whole course of the history of the Aryan race, and the rays of Regulus bring again the dreams of Babylon and Nineveh, of Greece and Rome, of India, and of the star-watching deserts of Arabia. Cyrus, in his conquering marches, may have looked to that star for help and inspiration, for it was the heavenly guardian of the Persian monarchs."

Regulus appears above the horizon a very little north of east about 9 P.M. on the evening of New Year's Day, and culminates at 9 P.M. April 6th.

The star β Leonis, or "Denebola," from an abbreviated Arab title meaning "the Lion's Tail," is an interesting star. It marked the tenth Arab lunar station known as "the Changer," *i. e.*, of the weather, and Al-Biruni wrote of it: "The heat turns away when it rises, and the cold turns away when it disappears." Denebola is one of the stars forming the so-called "Diamond of Virgo," a great diamond formed by the four stars Denebola, Arcturus, Cor Caroli, and Spica.

In astrology Denebola was considered unlucky, portending misfortune and disgrace. Its spectrum is Sirian, and it is approaching our system at the rate of about twelve miles a second. It is said to be thirty-three light years distant, and about ten times as bright as the sun. In all probability Denebola was a brighter star in former times than now, for Al-Sufi speaks of it as "the brilliant and great star of the first magnitude which is found on the tail." It comes to the meridian at 9 P.M. on the 3d of May.

γ Leonis, also called "Algeiba," an Arab name meaning "the Forehead," is one of the finest double stars in the heavens. Doberck estimates its period as four hundred years. Both stars can be seen very well in a three-inch telescope, with a power of 130, and the marked contrast of colours renders it a beautiful object. The colours are bright orange and greenish yellow. Sir Wm. Herschel discovered its duplicity in 1782. This star is approaching our system at the rate of twenty-four miles a second.

δ Leonis bears the Arab name "Zosma," the "Girdle." Ulugh Beg called it "Duhr," the "Lion's Back." In China it was known as "the High Minister of State." It is said to be approaching us at the rate of nine miles a second.

ϵ and μ Leonis were designated by an Arab writer as being "a whip's length apart." The distance is a little over two degrees.

π and σ Leonis were known to the Chinese as "the Honourable Lady" and "the Higher General" respectively.

Lepus
The Hare

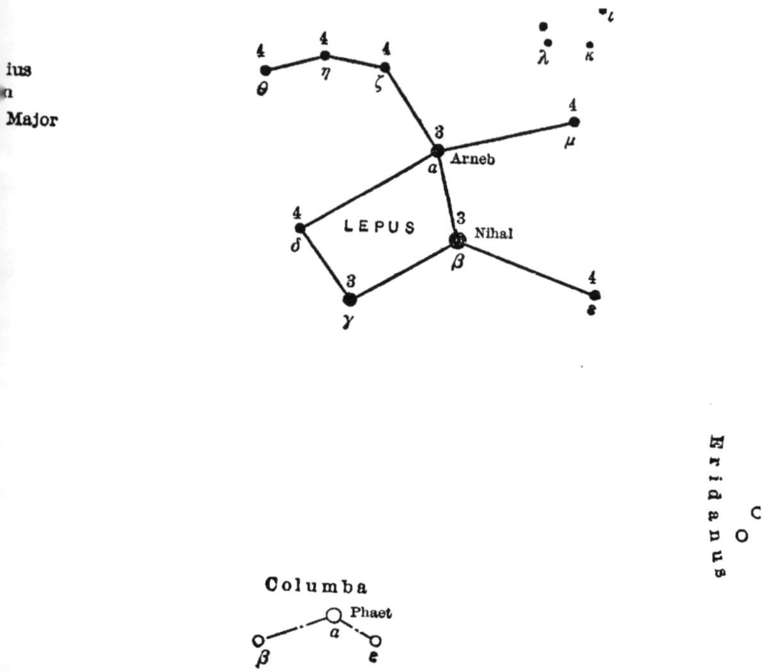

LEPUS

LEPUS
THE HARE

Under Orion's feet, mark too the Hare,
Perpetually pursued. Behind him Sirius
Drives as in chase, hard pressing when he rises,
And when he sinks as hotly pressing still.

FROTHINGHAM'S *Aratos.*

ONE of the chief characteristics of the constellation figures is the element of strife and conflict that seems to be especially emphasised, and is the predominating feature in many of the star groups. Thus the giant Hercules brandishes a club and tramples on the Dragon's head; Orion attacks the onrushing Bull; Ophiuchus struggles with a writhing Serpent, and crushes underfoot the Scorpion, which in turn thrusts at him with its sting; the Hounds, driven on by the Herdsman, continually pursue and harass the Great Bear; the fierce monster of the deep, the Whale, seems eagerly looking for whom he may devour, while the champion Perseus, with drawn sword, stands ever ready to join in mortal combat; the Archer aims his shaft at the heart of the Scorpion, the Hydra pursues the Lesser Dog, and is in turn in danger of being seized by the Crab; and here we find the timid Hare fleeing before the Hounds of Orion.

The story the stars unfold is therefore one replete with action and strife, and this fact is further evidence that the constellations were deliberately planned, for a haphazard arrangement of figures, passive in their attitudes, would savour of no special significance; but action calls for a plan and a definite idea that is preconceived, and so we find in the constellations an endeavour on the part of primitive man, through the medium of symbolism and allegory, to depict in the starry heavens for all ages, the predominant features of their lives, with special emphasis laid on the manifestations of nature, and the phenomena coincident with the creation of the world.

Classical writers are much in doubt as to the history of the constellation Lepus. It is situated directly south of Orion, and was one of the animals which the giant hunter is said to have delighted in hunting. It is for this reason, so it is said, placed near him among the stars.

Lepus is an inconspicuous constellation, and in these latitudes seems to crouch low on the horizon as if in an endeavour to escape attention.

According to Brown, the Hare is a reduplication of the moon, and as the sun seems to put to flight the moon, and as the solar overcomes the lunar light, so Orion pursues and conquers the Hare. An astonishing amount of folk-lore connects the moon and the Hare. Allen in his *Star Names and their Meanings* relates much that is of interest in this connection.

Dr. Seiss claims that in the Persian and Egyptian zodiacs the figure represented beneath the foot of Orion is not a Hare but a Serpent. If this is the case, we would have among the constellations the figures of three giants engaged in subduing serpents, surely sufficient to fully emphasise the enmity that instinctively exists between mankind and serpentkind. Schiller regarded Lepus as representing Gideon's Fleece.

Lepus does not rise until Aquila, the Eagle, the bird which loves the sun, is setting, from which fact arose the mythological belief of the hatred existing between the Hare and the Eagle.

As Lepus sets the Crow rises, and this fact accounts for the ancient belief that the Hare detested the voice of the Raven.

The early Arabs sometimes called this constellation "the Chair of the Giant" or "the Throne of Jauzah," owing to its position in the sky close beneath Orion. The Arabs also likened the four stars forming the quadrilateral which identifies the constellation, to four camels slaking their thirst in the near-by river in the sky, the Milky Way, or possibly Eridanus, the River Po.

Hewitt says that in early Egyptian astronomy Lepus was "the Boat of Osiris," the great god of that country, and identified with Orion.

The Chinese called the constellation "a shed." Lepus has been thought by some to represent certain Biblical figures such as "the Magdalen in tears," "Judas Iscariot," or "Cain driven from the face of the earth to the face of the moon."

α Leporis was called "Arneb" by the Arabs. It is a double star, the stars being coloured pale yellow and grey. It culminates at 9 P.M. on Jan. 24th.

Six seconds away from Alpha is situated Sir John Herschel's 3780, a sextuple star, a beautiful object even in a small telescope.

β Leporis, known to the Arabs as "Nihal," is a triple star of magnitudes 3d, 10th, and 11th.

Lepus contains the celebrated variable R. Leporis, of a deep crimson colour. It was discovered by Hind in 1845, and is sometimes referred to as "Hind's Crimson Star." It has been likened to "a drop of blood on a black field." No other star in these latitudes compares with it in depth of colour.

Just west of Lepus is the little asterism known as "the Brandenburg Sceptre," designed by Kirch in 1688. It contains but four stars of the 4th and 5th magnitudes, and the sceptre is represented in Burritt's Atlas as standing upright in the sky.

Libra
The Scales

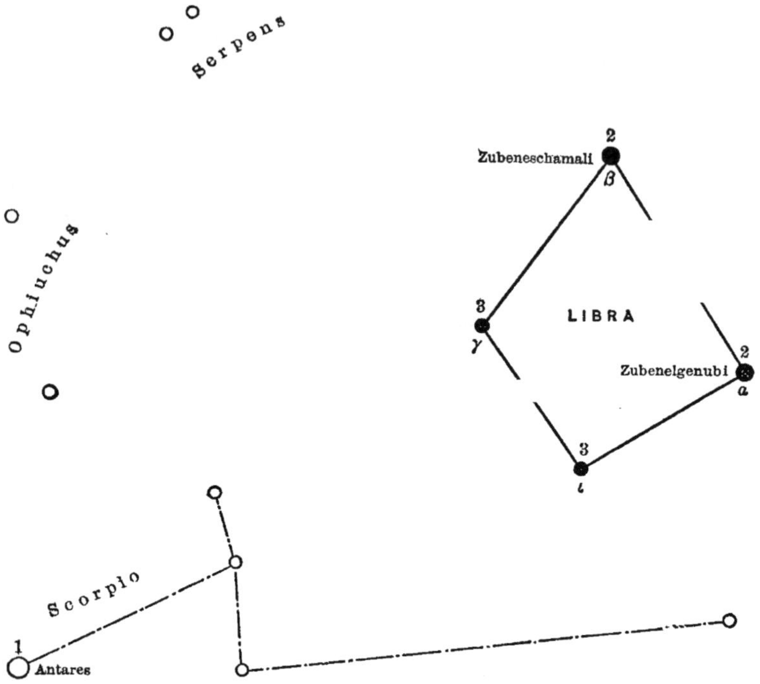
Serpens
Ophiuchus
2
Zubeneschamali
β
3
γ
LIBRA
2
Zubenelgenubi
α
3
ι
Scorpio
1
Antares

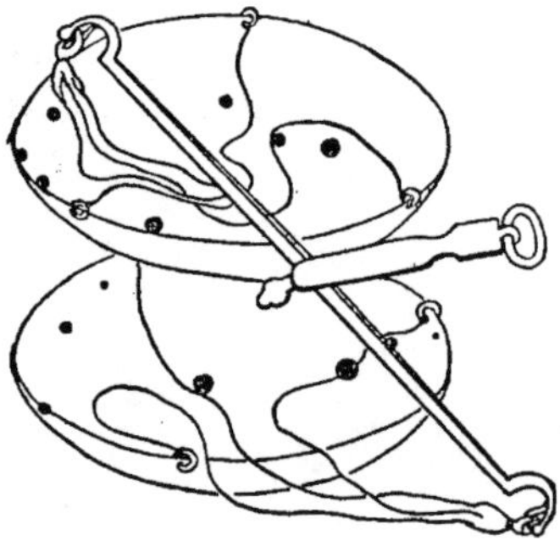

LIBRA

LIBRA
THE SCALES

Libra weighs in equal scales the year.

THOMSON.

OF the zodiacal signs Libra is the only one not Euphratean in its origin, the figure having been imported from Egypt. It is also the only zodiacal constellation that represents an inanimate object. It represented originally the balance of the sun at the horizon between the upper and under worlds, and secondarily the equality of the days and nights at the equinoxes.

The title "Libra," the "Balance," we owe to the Romans, but it is not known definitely how far back into antiquity the symbol goes. The constellation is anciently represented by the figure of a man holding a pair of scales. The human figure is omitted in all Arabian zodiacs, as it was held unlawful by the believers in the Koran to make any representations of the human form. On Burritt's Atlas also the Scales appear alone.

The Greeks combined this constellation with the Scorpion, and the stars in Libra formed the claws of the creature. Greek writers mention "Chelæ Scorpionis" (the claws of the Scorpion) in the place of Libra.

Libra seems to have been made an individual constellation, and separated from Scorpio, in the time of Julius Cæsar, for the Romans placed here the figure of Julius Cæsar holding a balance in his hand, instead of the Claws, and among the titles "Libra" was commonly employed. In after time, the figure of the Emperor was taken away, and the Scales only were retained as we now see them.

Some authorities hold that, in spite of the fact that the Greeks did not recognise a constellation figure between Virgo and Scorpio, an independent constellation existed at an earlier date. It is not clear just why the Greeks failed to discover it.

Serviss states that there are indistinct indications that in the valley of the Euphrates, the constellation now known as Libra stood for the Tower of Babel.

Besides the analogy mentioned respecting the sign of the Balance and the equality of the nights and days at the time of the autumnal equinox, we find that the Balance was the emblem of the office of Virgo, as the goddess of justice, so that there seems to have been a desire to connect these two constellations.

The Balance in poetical fiction belongs to the goddess Astræa, and in the pans the fate of mortals was supposed to be weighed.

The symbol of the sign Libra, ♎, represents it is said the beam of a pair of scales in equilibrium, thus denoting the equal duration of the nights and days. Brown however thinks that the symbol represents the top of the archaic Euphratean altar, located in the zodiac next preceding Scorpio, and figured on early gems, tablets, and boundary stones.

Allen points out that the stars in Libra, α, μ, ξ, δ, β, χ, ζ, and ν, seem to represent a circular altar. These stars were also thought to represent a censer, or a lamp and fire.

On an ancient zodiac there appears between the constellations of the Virgin and the Scales, the figure of a mound or altar, round which a serpent twines. Miss Clerke recalls the association of the seventh month, "Tashritu," with the seventh sign, and with the Holy Mound, Tul Ku, designating the Biblical Tower of Babel surmounted by an altar, so there is little doubt that in very early times the ancients saw in these stars an altar towering to the skies.

Libra is thus connected with the constellation Ara, the Altar, just south of it, and many have considered that in these figures are represented the altar of Noah, erected after the Deluge.

In Brown's Euphratean star list Libra is designated "the Claws," "the Life Maker of Heaven," and "the Lofty Altar."

Libra has been a great favourite with the poets of all ages. Manilius thus alludes to the Starry Balance:

> Then Day and Night are weigh'd in Libra's Scales
> Equal a while.

Milton refers to the constellation in his *Paradise Lost:*

> Th' Eternal to prevent such horrid fray
> Hung forth in heav'n his golden scales yet seen
> Betwixt Astræa and the Scorpion sign,
> Wherein all things created first he weighed.

And Homer sings:

> Th' Eternal Father hung
> His golden Scales aloft.

But Allen thinks this is not a reference to our Libra. Longfellow in his "Occultation of Orion" wrote:

> the scale of night
> Silently with the stars ascended.

And in his "Poet's Calendar" for September we read:

> I bear the Scales, when hang in equipoise
> The night and day.

In India Libra was regarded as a Balance, and in the zodiac of that country it is figured as a man bending on one knee and holding a pair of scales.

In China this constellation first represented a dragon, but afterwards a celestial Balance. In their early solar zodiac it was the Crocodile, or Dragon, the national emblem.

The early Hebrews also regarded Libra as a Scale-beam, as did the Egyptians, and it plainly appears as such on the Denderah planisphere. "The Libra of the zodiac," says Maurice, in his Indian Antiquities, "is perpetually seen upon all the hieroglyphics of Egypt, which is at once an argument of the great antiquity of this asterism, and of the probability of its having been originally fabricated by the astronomical sons of Misraim."

The beam was the instrument used by the Egyptians in measuring the inundations of the Nile, and some claim that because of this the beam was honoured by a place among the stars.

The Egyptian symbolic head-dress which appears in many representations of their ancient gods, as shown in the illustration, has, according to Plunket, an astronomical significance in which the constellation Libra figures. The two feathers represent the equal weights of the scale of Justice, and there also appear the horns of a goat and ram, and the disc enclosing a scarabæus, so that the head-dress is really an astronomic monogram containing four constellation figures in one, Aries, Cancer, Libra, and Capricornus.

According to Virgil the ancient husbandmen were wont to regard this sign as indicating the proper time for sowing their winter grain.

> But when Astræa's balance hung on high
> Betwixt the nights and days divides the sky,
> Then yoke your oxen, sow your winter grain,
> Till cold December comes with driving rain.

One of the early titles for Libra was Ζυγόν or Ζυγός, the Latin "Jugum," meaning the "yoke." This may have had reference to the yoking of the oxen mentioned in the poet's verses. The constellations were in general used as perpetual almanacs, and their seasons of appearance and disappearance were often warnings to the ancient tillers of the soil to sow their seed or reap their harvests.

Burritt tells us that the Balance was placed among the stars to perpetuate the memory of Mochis, the inventor of weights and measures. Those who refer the zodiacal constellations to the twelve tribes of Israel, ascribe the Balance to the tribe of Asher.

According to Serviss, Libra seems to be identical with a Mayan constellation, with which was associated a temple where dwelt a priest whose special business it was to administer justice, and to foretell the future by means of information obtained from the spirits of the dead.

The Peruvian asterism corresponding to Libra was entitled "Rainbow Lightning," "Sacred or Divided River," and "the Earth." These titles in a measure indicate the tempestuous nature of the weather when the sun was in this sign.

To the early Christians, Libra represented the Apostle Philip, and Cæsius identified it with the balances of the book of Daniel, in which Belshazzar had been weighed and found wanting.

In astrology Libra is the House of Venus. Those born from Sept. 23d to Oct. 23d are said to be ruled by this sign. The natives of this constellation are tall and well made, says Proctor, elegant in person, round faced, and ruddy but plain featured. When old they are of sweet disposition, just and upright in dealings. It governs the lumbar regions and reigns over Austria, Alsace, Savoy, Portugal, India, Ethiopia, Lisbon, Vienna, Frankfort, Antwerp, and Charleston.

It is a masculine sign and fortunate. Vulcan was the deity that watched over it. The significant flower was the violet, the precious stone the diamond.

The southern scale meant bad fortune, the northern on the contrary was eminently fortunate.

Only two of the stars in the constellation are specially interesting. α Libræ was called by the Arabs "Zubenelgenubi," meaning the southern claw. It is a wide double and culminates at 9 P.M. on the 17th of June. It marks the Hindu lunar station signifying "Branched."

It is a curious fact that the Peruvians of the western world had in connection with Libra a ceremonial purification by bathing at the junction of two streams. This would be where the stream *branched* out, which shows clearly a similarity of representation respecting this star between two widely separated peoples.

β Libræ was known by an Arab name signifying the northern claw. It is the only naked eye green-coloured star in the heavens, and is an interesting variable. Eratosthenes called it the brightest of all the stars in the Scorpion, that is in the double constellation, and Claudius Ptolemy gives it as equal with Antares, the brilliant first magnitude star in the heart of the Scorpion. As it is now a full magnitude fainter than Antares, it must have lost much of its pristine brilliance, though there is a possibility that Antares may have increased in brilliance. Beta has a Sirian spectrum, and is said to be approaching our system at the rate of six miles a second.

δ Libræ is a variable star of the Algol type, discovered by Schmidt in 1859, with a period of nearly two days and eight hours.

The constellation is easily identified as its four principal stars form a fairly conspicuous quadrilateral figure. About twenty-two centuries ago this constellation coincided with the sign Libra, but owing to the Precession of the Equinoxes it has advanced thirty degrees on the ecliptic, and the constellation Scorpio is now in the sign sagittarius and so on.

The Egyptian Symbolic Head-dress.

Lyra
The Lyre

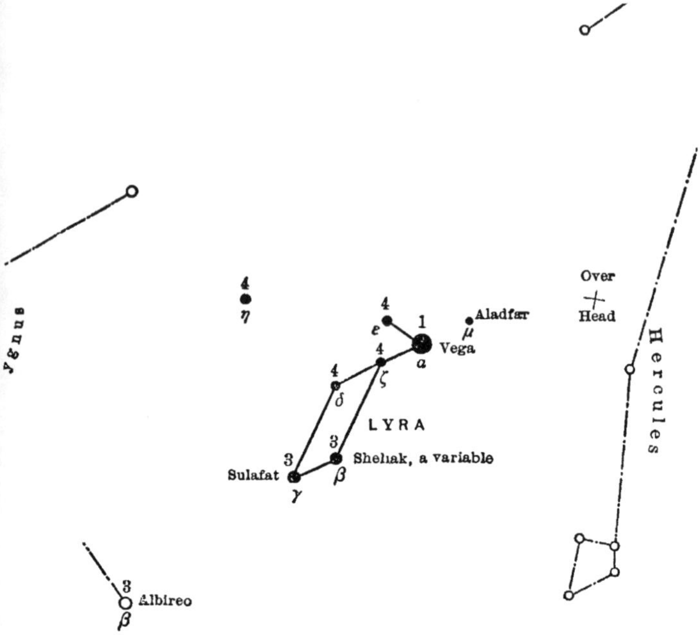
ygnus
4
η
4
ε
1
μ
Aladfar
Vega
α
4
ζ
4
δ
LYRA
3
Sheliak, a variable
3
Sulafat
β
γ
Over
Head
Hercules
3
Albireo
β

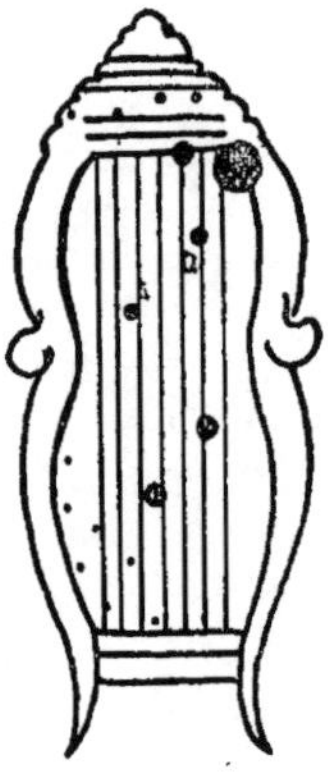

LYRA

LYRA
THE LYRE

> The Lyre whose strings give music audible
> To holy ears.
>
> LOWELL.

IN mythology Lyra is the celestial harp invented by Hermes, which Apollo or Mercury gave to Orpheus, the skilled musician of the Argonautic expedition.

There are many references among the poets to the wonderful talent of this harpist. Shakespeare says of him:

> Everything that heard him play,
> Even the billows of the sea
> Hung their heads and then lay by.

And again in the *Two Gentlemen of Verona* we read:

> For Orpheus' lute was strung with poet's sinews;
> Whose golden touch could soften steel and stone,
> Made tigers tame, and huge leviathans
> Forsake unsounded deeps to dance on sands.

It is related that Orpheus even descended to the infernal regions, and charmed Pluto, the King of Hell, with the music of his harp, so that he won from Pluto his lost bride, Eurydice; but as the legend goes, lost her again, by looking backward which he had been forbidden to do as he emerged from Hades. After his death he received divine honours, and his lyre became one of the constellations.

Max Müller identifies Orpheus with the Sanscrit "Arbhu," used as a title for the sun. According to this explanation, the sun follows Eurydice, "the wide-spreading flush of the dawn, who has been stung by the serpent of night," into the regions of darkness. There he recovers Eurydice, but while he looks back upon her she fades before his gaze, as the mists of the morning vanish before the glory of the rising sun.

Cox found in the music of Orpheus the delicious strains of the breezes which accompany sunrise and sunset.

Mrs. Martin thus delightfully refers to Lyra: "It is easy to get some sense of the fancy that gave the constellation its name as we watch it during the lovely spring evenings, floating lightly in the sky, the parallel lines connecting its principal stars vaguely suggestive to the willing mind of some quaint stringed instrument that under a magic touch might send out heavenly music through the resonant air."

Lyra has also borne the title "the harp of Arion," Arion being a famous musician of the court of Periander, King of Corinth. The fable relates that, returning from Sicily, he was about to be thrown overboard by the sailors, when he requested permission to play his harp. This request being granted, presently dolphins appeared, enchanted by the sweet strains, and when Arion plunged into the sea, the dolphins, so it is said, bore him safe to land.

Brown tells us that the Hellenic myth connected with Lyra is the comparatively late story of Hermes (the Lord of Cloud) as the inventor of the Lyre from the tortoise, which is related in the Homeric Hymn.

The earlier history of the constellation is twofold, Euphratean and Phœnician. In the valley of the Euphrates it was originally one of the three birds opposed to Herakles. Thus its principal star, Vega, was known as "the Falling Grype." According to an Arab commentator on Ulugh Beg, ε and ζ Lyræ represent the two wings of the Grype, by drawing in which he let himself down to the earth.

On the Phœnician side Lyra is a musical instrument. Aratos names it "Xelus" (the little tortoise or shell), thus going back, says Allen, to the legendary origin of the instrument, from the empty covering of the creature cast

Photo by Anderson

Orpheus and Eurydice
Villa Albani, Rome

upon the shore with the dried tendons stretched across it.

Blake offers the following explanation of the connection of this figure with the tortoise: "At the probable time when the name of the constellation was composed, and the figure invented, Vega, the chief star in the constellation, may have been very near the Pole of the heavens, and therefore have had a slow motion, and hence it might have been named 'the Tortoise,' and this in Greek would easily be interpreted into Lyre."

This double meaning of the word seems certainly to have given rise to the fable of Mercury having constructed a lyre out of the back of a tortoise.

There was also a notion that the Lyre was placed in the sky near Hercules for the alleviation of his toil. There is the following interesting note on the Lyre by Burritt:

"The lyre was a famous stringed instrument much used by the ancients, said to have been invented by Mercury about the year of the world 2000, though some ascribe the invention to Jubal. It is universally allowed that the lyre was the first instrument of the stringed kind used in Greece. The different lyres at various periods of time had from four to eighteen strings each. The modern lyre is the Welsh harp. The lyre among painters is an attribute of Apollo and the Muses."

Emphasis seems to be laid on the mystic number seven in this constellation, as in the stars of Ursa Major, and the Pleiades, for the Lyre was mentioned by Ovid as having seven strings. Our Longfellow thus sings of it:

> I saw with its celestial keys
> Its chords of air, its frets of fire,
> The Samian's great Æolian Lyre,
> Rising through all the sevenfold bars,
> From earth unto the fixéd stars.

In Bohemia our Lyre was "the Fiddle in the Sky." The ancient Britons called it "King Arthur's Harp," and the

Persians "a Lyre." Novidius said it was "King David's Harp," and Schiller curiously enough thought that the constellation represented "the Manger," the birthplace of the infant Saviour.

Allen says that the association of Lyra's stars with a bird perhaps originated from a conception of the figure current for millenniums in ancient India, that of an eagle or vulture.

The Arabs called Lyra "the Swooping Eagle," to distinguish it from Aquila, which was regarded as "the Flying Eagle." Lyra has also been likened to a "Goose," an "Osprey," a "Wood Falcon," and a "Kite." The Hindus figured the stars α, ε, and ζ Lyræ as a triangle, or as the three-cornered nut of an aquatic plant.

Notwithstanding the singularly diverse ideas as to the figures represented by this star group, the name generally applied to it has been "Lyra," and the figure so shown from ancient times. Roman coins still in existence show it thus. According to Dr. Seiss, Lyra symbolises the rejoicing in heaven at the final victory over the powers of evil. To the early Christians Lyra represented the Saviour's Manger, and David's Harp.

From this constellation radiate the swift meteors known as "the Lyrids." The maximum of the shower is on the 19th and 20th of April.

Lyra is noted because of its lucida, the brilliant Vega, "the glory of the summer heavens."

The poet thus sings of Lyra and Vega:

> One of these illuminates
> The heavens far around, blazing imperial
> In the first order.

The Arabs called Vega "the Falling Vulture." It has also been called "the Harp Star," and "the Arc-light of the Sky."

It has a decided bluish tint, and is one of the most beautiful stars in the northern hemisphere.

Photo by Anderson

Mercury, by Rubens
Gallery of the Prado, Madrid

Manilius, who wrote in the age of Augustus, thus alludes to Vega:

> One, placed in front above the rest, displays
> A vigorous light and darts surprising rays.

Among Latin writers Vega was called "Lyra" in classical days.

> Azure Lyra, like a woman's eye
> Burning with soft blue lustre.
>
> Willis.

The Romans made much of this star, for the beginning of their autumn was indicated by its morning setting. Brown writes of it:

"At one time Vega was the Pole Star, and known to the Akkadai as 'the Life of Heaven,' and to the Assyrians as 'the Judge of Heaven.'"

The Chinese and Japanese call Vega "the Spinning Maiden," or "the Girl with a Shuttle." She was supposed to stand at one end of the magpie bridge, over the Milky Way, awaiting her lover. This legend was related in connection with the history of the constellation Aquila.

Lockyer claims that some of the temples at Denderah in Egypt were oriented to Vega as early as 7000 B.C.

Owing to the phenomena of Precession, Vega will be the Pole Star 11,500 years hence.

It is almost in a direct line towards this blazing blue sun that the solar system is flying through space at the rate of twelve or fifteen miles a second. This goal of our sun and its family of planets is known as "the Apex of the Sun's Way." The accompanying diagram indicates its location according to different authorities. See p. 263.

Vega is the second brightest star to be seen in this latitude, Sirius alone surpassing it in splendour. In spite of its great brilliance, Vega is not one of our near neighbours. According to Peck it is eighteen light years distant, some authorities say twenty-nine. If the distance from the earth to the sun is regarded as one foot, that from Vega would be

158 miles, and if our sun occupied the place of Vega, it would appear to us as a faint point of light just visible to the naked eye. Vega is said to surpass our sun in brilliance a hundredfold, and is approaching our system at the rate of 9.5 miles a second.

"It is a young orb," says Serviss, "blazing with the white fire of stellar youth, dazzling the eye with the strange splendour of its gem-like rays, which possess the piercing quality of the reflections from a blue-white diamond."

Mrs. Martin pays the following tribute to this azure-tinted sun:

"About three hours after Arcturus has risen there will come peeping over the north-eastern horizon a brilliant, bluish star which twinkles so gaily and commands such instant admiration that its entrance into view has almost a dramatic effect. This is Vega, the third of the trio of bright stars that give a May-dance around the pole. . . . Early in May the star rises at about the same hour that the sun sets, and all summer long it is the gayest and perhaps the most instantly attractive star in the evening skies. . . . Vega has a companion star, much smaller than itself, revolving around it, which is of the same beautiful bluish colour as the larger star. The companion is of about the tenth magnitude and can be seen only with a large telescope. Vega is about four thousand times brighter than her companion."

Vega is visible at some hour of every clear night throughout the year and culminates at 9 P.M., Aug. 12th.

β Lyræ, known to the Arabs as "Sheliak," is a noted variable. Goodricke in 1784 was the first to detect changes in its brilliancy, and Argelander carefully observed the star for nineteen years, 1840 to 1859. Its period is 12 days 21¾ hours, though it has remarkable and unexplained variations in light. Scheiner says of it, "There is great probability that more than two bodies are concerned in the case of β Lyræ." This star is one of ten that are said to be pear-shaped, a fact that may account for its light variations.

Between β and γ Lyræ is the wonderful "Ring Nebula." It is the only annular nebula visible through small telescopes.

ε Lyræ is the celebrated "double double star," a star almost a naked eye double, and each of these stars is in turn double. A three-inch telescope with a power of 130 will separate these stars.

γ Lyræ, 2½° east of β was known as "Sulafat," one of the early titles of the constellation. Another name for it was "Jugum." It is a bright yellow star of the 3.3 magnitude.

The remaining stars in the constellation are of no special interest.

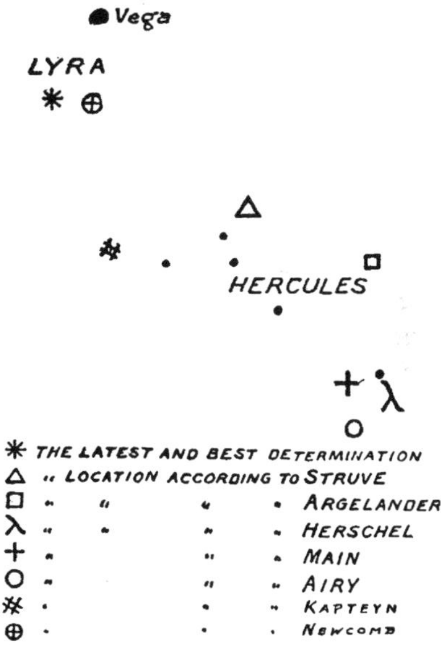

APEX OF THE SOLAR SYSTEM

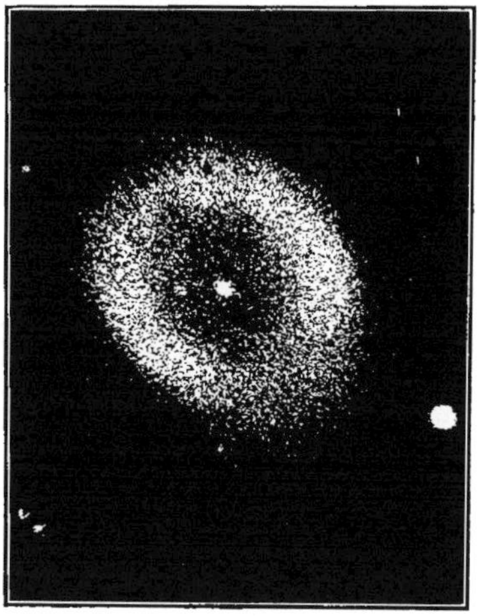

Ring Nebula in Lyra

Ophiuchus or Serpentarius, the Serpent Bearer and Serpens, the Serpent

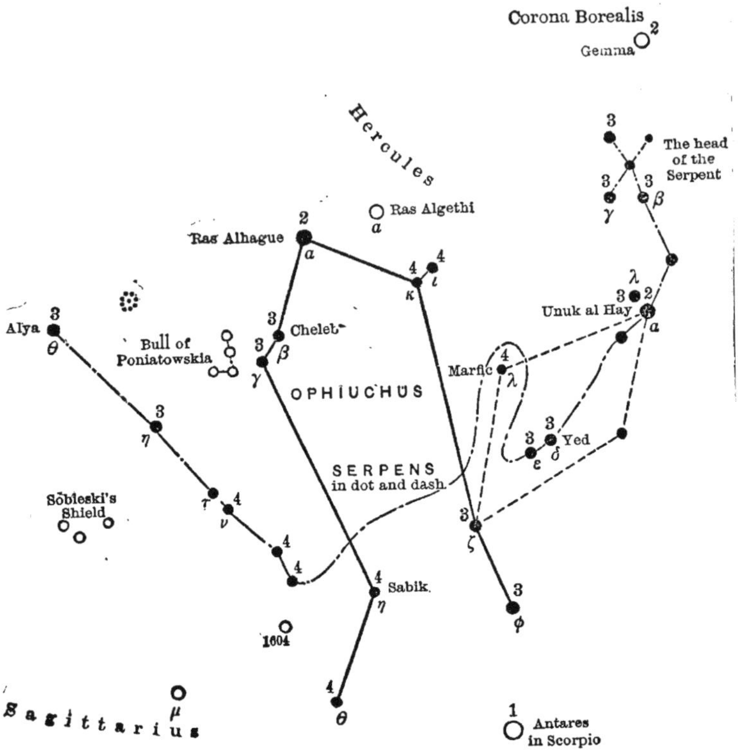
Corona Borealis
Gemma
Hercules
The head of the Serpent
Ras Algethi
Ras Alhague
Unuk al Hay
Alya
Bull of Poniatowskia
Chelet
OPHIUCHUS
Marfic
Yed
SERPENS in dot and dash
Sobieski's Shield
Sabik.
1604
Sagittarius
Antares in Scorpio

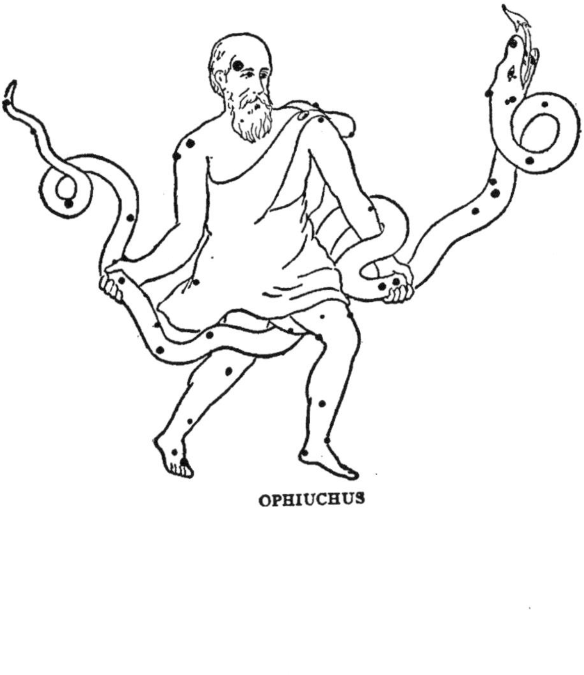

OPHIUCHUS

OPHIUCHUS OR SERPENTARIUS, THE SERPENT-BEARER, AND SERPENS, THE SERPENT

Thee, Serpentarius, we behold distinct,
With seventy-four refulgent stars.

EUDOSIA.

THE title "Ophiuchus" is derived from the Greek words ὄφι and οὔχος, meaning the man that holds the serpent. According to Plunket the constellation was probably invented about 3500 B.C. in latitude 35° north. "At this time the constellation would have been in opposition to the sun at the season of the spring equinox, triumphing over the powers of darkness, namely the Scorpion, on which the giant Ophiuchus treads, and the Serpent which he crushes in his hands."

Of all the constellational figures, Hercules and Ophiuchus are perhaps the most remarkable in conception and design, for they are each clearly intended to be combined with star groups which from time immemorial have been universally considered to represent serpents. They are the only two constellations that seem identical in design, and in these ancient constellations we see clearly a desire on the part of primitive man to place on record, for all the ages to read, the great fact of the triumph of man over the serpent, the symbol of the powers of evil.

Krishna, one of the most revered gods of India, is often represented as standing with one foot on a serpent's head, and holding it up by the tail, a reduplication of the figures and the ideas embodied in the constellation figures, Hercules and Ophiuchus.

On the old maps, Ophiuchus is represented as a venerable

man, having both hands clenched in the folds of a great serpent which is writhing in his grasp with its head close to the Crown; whence the Serpent is often said "to be licking the Crown." The constellation is of great antiquity, as the records show that it was known to the ancients twelve hundred years before the Christian era. Homer refers to it, and Aratos clearly describes the figure in the following lines:

His feet stamp Scorpion down, enormous beast,
Crushing the monster's eye and platted breast.
With outstretched arms he holds the Serpent's coils,
His limbs it folds within its scaly toils,
With his right hand its writhing tail he grasps,
Its swelling neck his left securely clasps,
The reptile rears its crested head on high,
Reaching the seven-starred Crown in Northern sky.

There is something remarkable in the central position of this constellation. It is situated almost exactly in the mid-heavens, being nearly equidistant from the Poles, and midway between the vernal and autumnal equinoxes. The commanding location of the constellation makes the figure and its intended representation especially significant.

Manilius thus refers to the Serpent-Bearer:

Next Ophiuchus strides the mighty snake,
Untwists his winding folds and smooths his back,
Extends his bulk, and o'er the slippery scale
His wide stretched hands on either side prevail.
The snake turns back his head and seems to rage
That war must last where equal power prevails.

In Greek mythology Ophiuchus was the great physician Æsculapius, with whose worship serpents were always associated, as symbols of prudence, wisdom, renovation, and the power of discovering herbs, and the constellation was often called "Æsculapius" or "the god of medicine."

Æsculapius was said to have been educated by his father Apollo, or by the centaur Chiron, and was the earliest of his profession, and accompanied the Argonautic expedition. Afterwards he became so skilled in practice that it is said he even restored the dead to life. His success in this latter achievement so alarmed Pluto that he persuaded Zeus to remove Æsculapius to the sky.

One of the last acts of Socrates was to offer a cock to Æsculapius, and the cock and serpent were ever sacred to this great physician. He was worshipped at Epidaurus, a city of Peloponnesus, and hence he is styled by Milton, "the god in Epidaurus."

In his *Paradise Lost*, Milton thus refers to Ophiuchus:

> the length of Ophiuchus huge.

In the Middle Ages, the Serpent-Bearer was sometimes regarded as symbolising Moses with the Brazen Serpent, and Golius insisted that this sky figure represented a Serpent Charmer. Al-Sufi's title, "le Psylle," meaning one skilled in the cure of snake bites, seems to confirm this view.

Ophiuchus is also identified with Laocoön, the priest of Neptune, who during the siege of Troy was attacked and strangled by sea serpents, for his irreverent treatment of the wooden horse.

Pliny regarded the stars in this constellation as dangerous to mankind, occasioning much mortality by poisoning.

The Serpent-Bearer has also been thought to represent Saint Paul, with the Maltese viper, Aaron, whose staff became a serpent, Saint Benedict, and the Great Physician.

The constellation is noted for the number of new stars (*novæ*) which have appeared within its borders,—one in 1230, "Kepler's Star" in 1604, and one in 1848. It would seem as if this part of the sky should be especially observed.

Hill calls attention to the fact, that, although Ophiuchus is not one of the zodiacal constellations, yet out of the twenty-five days from Nov. 21st to Dec. 16th, which the sun spends in passing from Libra to Sagittarius, only nine are spent in the Scorpion, the other sixteen being occupied in its journey through Ophiuchus.

Ophiuchus contains the discarded asterism known as "the Bull of Poniatowski," the Polish Bull. It consists of but four stars, three of the fourth magnitude, and one of the fifth, situated about fifteen minutes east of the star γ Ophiuchi. One of the stars in this group (70) is an interesting binary with a period of about eighty-eight years. This star is estimated to be 120 quadrillions of miles away. Prey finds that the fainter of the two stars has four times the mass of the brighter star.

α Ophiuchi, known to the Arabs as "Ras Alhague," meaning the "Head of the Serpent Charmer," is a second magnitude star six degrees east of α Herculis. In China it was known as "How," the "Duke." It is said to be receding from the earth at the rate of twelve miles a second, and culminates at 9 P.M. July 28th.

β Ophiuchi was named "Cebalrai," or "Cheleb," meaning "the Heart of the Shepherd," the stars α Ophiuchi and α Herculis representing the shepherd and his dog.

Of the four reptiles that are found among the constellation figures, Serpens is the Serpent. Statius thus writes of it:

> Vast as the starry serpent that on high
> Tracks the clear ether and divides the sky,
> And southward winding from the northern Wain
> Shoots to remoter spheres its glittering train.

Burritt tells us that the Hivites of the Old Testament were worshippers of the Serpent, and that this idolatry was extremely ancient. Serpens was identified with Eve's temple in the Garden of Eden, and in the astronomy of Arabia it was known as "the Snake," although before the Hellenic influence was felt in Arabia, the stars in this region of the heaven were regarded by the Arabs as representing a pasture.

The Hebrews knew this star group as "the Serpent,"

Photo by Anderson

Laocoön
Museum of Vatican, Rome

from the earliest times. The space between ν and ε Serpentis was called by the Chinese "the Enclosure of the Heavenly Market."

The head of the Serpent is represented by an "X"-shaped group of stars, just south of the Northern Crown, which serves to identify the figure.

None of the stars in the Serpent is of special interest.

Orion
The Giant Hunter

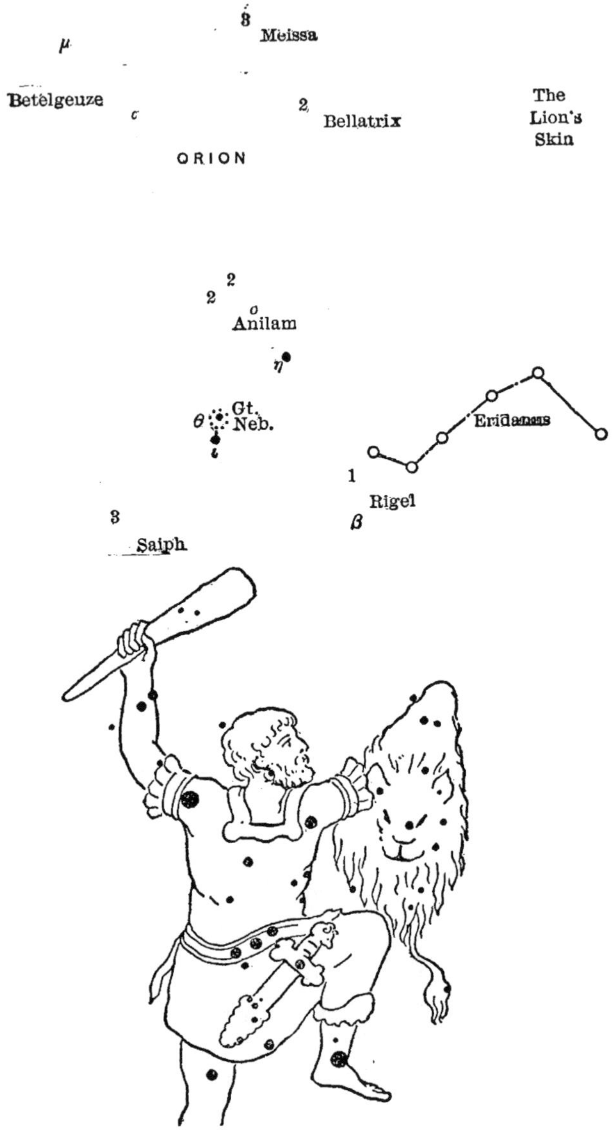
Meissa
Betelgeuze
Bellatrix
The Lion's Skin
ORION
Anilam
Gt. Neb.
Eridanus
Rigel
Saiph

ORION

ORION
THE GIANT HUNTER

Orion kneeling in his starry niche.

LOWELL.

THE constellation Orion has been the admiration of all ages, and vies with Ursa Major and the famous Pleiades in historical and mythological interest. It is beyond question the most brilliant of the constellations, containing as it does two stars of the first magnitude, and four of the second. With the exception of the "Dipper," the so-called "Belt of Orion" is probably the best known and most popular of all stellar objects.

The constellation is visible from every part of the globe, and the poets of all nations have sung its praises. Manilius pays the following tribute to the mighty hunter:

Now near the twins behold Orion rise,
His arms extended measure half the skies:
His stride no less. Onward with steady face,
He treads the boundless realms of starry space,
On each broad shoulder a bright gem displayed
While three obliquely grace his mighty blade.

And again he sings:

Orion's beams, Orion's beams:
His star gemmed belt and shining blade
His isles of light, his silver streams,
And glowing gulfs of mystic shade.

Shelley in his *Revolt of Islam* wrote:

While far Orion o'er the waves did walk
That flow among the isles.

Lucy Larcom contributes:

Orion with his glittering belt and sword
Gilded since time has been, while time shall be.

And Longfellow thus alludes to this beautiful constellation:

Begirt with many a blazing star
Stood the great giant Algebar
Orion, hunter of the beast,
His sword hung gleaming by his side.

Hesiod wrote:

When strong Orion chases to the deep the Virgin stars.

Tennyson refers to the constellation in his *Locksley Hall, Maud,* and *The Princess,* and Spenser describes the setting of Orion in these words:

And now in ocean deep
Orion flying fast from hissing snake
His flaming head did hasten for to steep.

Much doubt and mystery surround the title of the constellation. Brown, one of the most reliable authorities, is of the opinion that it is from "Uru-anna," meaning the "light of heaven," and that the title originated in the Euphratean Valley.

It seems reasonably certain that a star group of such prominence should have attracted attention from the earliest times, and that this constellation therefore is of great antiquity.

Maunder tells us that the word from which Orion was derived was "Kĕsil," a word which occurs in an astronomical sense four times in the Bible. The Hebrew word "Kĕsil" signifies "a fool," meaning a godless and impious person. In the Scriptures this word is associated with a word which, translated, refers to the Pleiades, sometimes likened to a flock of doves.

We have, therefore, in the figure of Orion, a mighty giant represented as trampling on a timid hare, and pursuing a flock of inoffensive doves, certainly a strange and incongruous association of figures.

Maunder claims that it was intense irony for the Hebrews to designate as "a fool" the constellation that the Babylonians had deified, and made their supreme god, and styled "the Mighty Hunter."

Orion has been identified with Merodach, probably the first King of Babylonia, and with the Nimrod of the Scriptures, "the mighty hunter before the Lord," known also as "the mighty one in the earth," a variant of Merodach.

Several Assyriologists consider that the constellations Orion and Cetus represent the struggle between Merodach and Tiamat. In support of this view, it may be said that Tiamat is expressly identified on a Babylonian tablet with a constellation near the ecliptic.

Maunder thinks that the view that has come down to us through the Greeks concerning Orion agrees much better with the associations of the constellations as held among the Hebrews rather than amongst the Babylonians. According to the Greek legend, Orion pursued the Pleiades, which were considered doves or virgins, and was confronted by the Bull. Cetus was not involved in the struggle, but was engaged in a combat with Perseus.

There is in the threatening attitude of the Mighty Hunter as he stands facing the advancing Bull, carrying

> . . . on his arm the lion's hide,

to ward off an attack, and his club raised to strike a blow, every indication of an attempt on the part of the inventor of the constellation to indicate a conflict to the death between Orion and the Bull.

The figure of the Hare crouching beneath the Hunter's foot is also significant. The hare has always been associated in folk-lore with the moon, and as Brown points out, Orion as "the light of heaven" is clearly identified with the sun. Here we have, many think, a figure symbolical of the perpetual strife between the powers of light and darkness, in which the former ever prevails.

Plunket claims 4667 B.C. as the date of the invention of this constellation. Orion at that time accurately marked the equinoctial colure, but others have thought that 2000 B.C. was the more probable date, as at that time the so-called "Belt of Orion" began to be visible before dawn in the month of June, called "Tammuz," and Orion was known to the Chaldeans as "Tammuz."

As the death of Adonis is celebrated in the month Tammuz, Miss Clerke is of the opinion that Orion received this name because its annual emergence from the solar beams coincided with the mystical mourning for the vernal sun. "Altogether the evidence is strong," says Miss Clerke, "that Orion may be considered as a variant of Adonis, imported into Greece from the East at an early date, and there associated with the identical group of stars which commemorated to the Akkads of old, the fate of Tammuz, the 'only Son of Heaven.'"

Homer describes Orion as the "tallest and most beautiful man," which description well befits Adonis.

According to Brown, Orion like Boötes was regarded as a shepherd, the keeper of the flock of stars, and one of his titles was "Shepherd Spirit of Heaven."

Orion was also known as "the Lord of the River Bank," an appropriate name as regards his location close by Eridanus, the River Po.

In mythology Orion is connected with the constellation of the Scorpion, and it is related that Orion boasted that there was not an animal on earth which he could not conquer. To punish his vanity, it is said, a scorpion sprang out of the earth and bit his foot, causing his death. At the request of Diana he was placed among the stars opposite his slayer.

Ovid agrees with this version, but Hyginus, Homer, and Apollodorus claim that Orion was killed by Diana's darts, and that he was placed in the sky opposite the Scorpion so

Photo 1

that "he might escape in the West as the reptile rose in the East."

When the Scorpion comes
Orion flees to utmost end of earth.

The Hindus connect in a legend, Aldebaran, the red star in the eye of the Bull, Sirius, known to them as "the Deer-slayer," and Orion, which they regard as "a Stag." The story is as follows: "The Lord of created beings fell in love with his daughter. She took the form of a dove and fled. He thereupon changed himself into a stag and pursued her, but was shot by Sirius, who was selected by the indignant gods to slay him."

The three stars in the head of the Mighty Hunter constitute one of the Hindu lunar stations known as "the antelope's head," in accordance with this myth.

Another legend concerning Orion relates that he was the lover of Merope, daughter of Œnopion, King of Chios. His suit was frowned upon, so he attempted to elope with the fair object of his affections. The King, however, discovered his perfidy, and drugging him, put out his eyes, and left him alone on the seashore. Following the sound of a hammer, Orion, it is said, made his way to the forge of Vulcan, where he besought assistance. Vulcan placed him on the shoulders of a Cyclops, who carried him to the top of a mountain, where, facing the rising sun, he received his sight.

. . . he
Reeled as of yore beside the sea
When blinded by Œnopion.
He sought the blacksmith at his forge
And climbing up the narrow gorge
Fixed his blank eyes upon the sun.

Longfellow.

This legend connects Orion with the Sun-god, and the title he sometimes bears, "Light of Heaven."

There is an analogous myth of the moon-goddess connected with Orion: The moon-goddess fell in love with the Giant Hunter. The sun-god did not approve of him, and resolved to bring about his destruction. As Orion was bathing, the sun-god poured his golden rays upon him, and called on the moon-goddess to test her skill in archery by shooting at the gleaming mark. The moon-goddess winged a shaft, and slew Orion, her lover, hidden in the brilliant light. Distracted she appealed to Jove, who placed Orion in the sky so that the moon-goddess might gaze upon him as she sails in her silver chariot.

Still another story relates that Orion was born like Athena without a mother, and became a famous iron worker, so skillful that Vulcan employed him to build a palace under the sea.

Orion was always regarded as a stormy constellation from the fact of its setting in the late autumn. Thus Æneas accounts for the storm which cast him on the African coast, on his way to Italy:

> To that blest shore we steer'd our destined way,
> When suddenly dire Orion rous'd the sea.

Again we read:

> Tell him that charg'd with deluges of rain
> Orion rages on the wintry main.

The constellation's stormy character, says Allen, appeared in early Hindu, and perhaps even in earlier Euphratean days, and is seen everywhere among classical writers, with allusions to its direful influence.

Polybios, the Greek historian of the second century before Christ, attributed the loss of the Roman squadron in the first Punic War to its having sailed just after "the rising of Orion."

Hesiod long before wrote of the same rising:

> then the winds war loud,
> And veil the ocean with a sable cloud.

And Milton wrote:

Photo by Anderson

Diana
Capitoline Museum, Rome

> When with fierce winds Orion arrived
> Hath vexed the Red Sea coast.

Hesiod also lays it down that the rising of Orion is the season for threshing:

> Forget not when Orion first appears
> To make your servants thresh the sacred ears;

and points to the time when Orion is in the mid-heavens as proper for the vintage. He also directs the husbandmen to plough at the setting of this constellation, and warns navigators to avoid the dangers of the sea, when the Pleiades, flying from Orion, are lost in the waves.

The Syrians and Arabians knew Orion as "the Giant." To the early Arabs, Orion was "Al-Jauzah," often erroneously translated "Giant," says Allen. Originally this was the term used for a black sheep with a white spot in the middle of the body, and this may have become the designation for the middle figure of the heavens, which, from its pre-eminent brilliancy, always has been a centre of attraction.

In Egypt, the soul of Osiris was said to rest in the constellation Orion, and in the round zodiac of the temple of Denderah there is a mythological figure of a cow in a boat identified as Sirius, and near it another mythological figure which has been proved, according to Lockyer, to represent the constellation Orion.

Allen says that the Egyptians represented Orion as Horus, the young or rising sun, in a boat surmounted by stars, and as "Sahu" in the great Ramesseum of Thebes, about 3285 B.C.

Orion was an extremely important constellation in Egypt, because it preceded and announced the approaching rise of Sirius, which in turn heralded the inundation of the Nile.

According to a Jewish tradition, this constellation was appropriated to himself by a particularly mighty man. The Hebrews knew Orion as "the Giant," bound to the sky for rebellion against Jehovah, and Allen thinks that this may be the explanation of the well-known phrase "the Bands or Bonds of Orion."

The Chinese called Orion "Tsan," which signifies "Three," and corresponds to the "Three Kings," a title sometimes applied to the three prominent stars in the "Belt." The Chinese also knew Orion as "the White Tiger," a title taken from the constellation Taurus, close to Orion.

The Eskimos called Orion's Belt "Tua Tsan," a title similar to the Chinese title, which might indicate that the Eskimos originally came from China as has often been contended.

The Eskimos thought that Orion represented a party of bear-hunters, with their sledge, and the bear they were pursuing, transported to the sky.

According to Dr. Seiss, Orion stands as a prophetic representation of the great enemy and destroyer, death.

The early inhabitants of Ireland called Orion "the armed King," and the Mayas, the ancient inhabitants of Yucatan, knew the constellation as "a Warrior," a further instance of a similarity in stellar nomenclature among widely separated nations, a similarity that is so marked and so often encountered as to disprove any idea of mere coincidence.

"The rising of Orion is one of the most imposing spectaeles that the heavens afford," says Serviss. "No constellation compares with it in brilliance. It is wonderfully rich in telescopic objects of interest, and Flammarion calls it 'the California of the Sky.'"

Mrs. Martin says the group exacts more immediate admiration because the bright stars are "clustered so closely and symmetrically as to form a set figure of dazzling jewels, a veritable sunburst of diamonds in the sky."

There is much of interest concerning the individual stars in this constellation. α Orionis was known to the Arabs as "Betelgeuze," an abbreviation for "the armpit of the

Photo by Mansell

The Zodiac of Denderah

central one." It is an irregular variable star of a rich topaz hue, and is often called "the Martial Star."

> . . . First in rank
> The martial star upon the shoulder flames.

In astrology this star denoted military or civic honours.

Mrs. Martin describes Betelgeuze as "suggestive of sombreness in its dull and comparatively untwinkling face."

Allen tells us that the title "Roarer" or "Announcer" is also applied to this star, as heralding the rising of its companions.

Betelgeuze marks the 6th Hindu lunar station known as "Ardra," meaning "moist." In this title we see an allusion to the stormy character of the constellation, and when this star rose the rainy season set in.

Sayce and Bosanquet identify Betelgeuze with the Euphratean "Gula," and Brown says the constellation of "the King" or "Ungal" refers to α, γ, and λ Orionis. In the Euphratean star list we find Betelgeuze styled "Lugal" (the King). The similarity in these titles "Gula," "Ungal" and "Lugal" is strikingly suggestive.

Secchi makes Betelgeuze a typical star of his third class with banded spectra, suggesting that it may be approaching the point of extinction. According to Vogel it is receding from our system at the rate of 10.5 miles a second, and culminates at 9 P.M. Jan. 29th.

β Orionis is known to us as "Rigel," the Arab title from which it came meaning "the left leg of the Jauzah, or Giant." Another name for it is "Algebar," a corruption of "Al-Jabbah," the "mighty one." It is a brilliant white star, and ranks fifth in order of brightness of all the stars visible in this latitude.

In astrology Rigel denotes splendours and honours.

In the Norseland, Rigel marked out the great toe of Orwandil, the other toe having been broken off by the god Thor, when frost-bitten, and thrown to the northern sky, where it became the little star Alcor in the handle of "the Dipper."

Rigel is receding from our system at the rate of ten miles a second. Newcomb estimated that it exceeds our sun in brilliance not less than ten thousand times. It is a double star, the companion being of the ninth magnitude, and blue in colour, but it is difficult to glimpse in a small telescope owing to the lustre of its primary.

γ Orionis is known as "Bellatrix," the "Female Warrior," and "the Amazon Star." It is pale yellow in colour, and of the second magnitude. One Arab title for it was "the Roaring Conqueror," or "the Conquering Lion." It marks the left shoulder of the Giant.

Allen tells us that in an Amazon River myth, Bellatrix figures as a young boy in a canoe with an old man, represented by the star Betelgeuze. They are said to be chasing the Peixie Boi, a dark spot in the sky near Orion.

In astrology Bellatrix was the natal star of all destined to great civil or military honours, and rendered all women born under its influence lucky and loquacious.

It is said to be receding from our system at the rate of five miles a second.

The three so-called "Belt Stars," of the second magnitude, in the centre of the parallelogram which renders the constellation conspicuous, have excited the attention of all ages, and many have been the titles bestowed on them. They are δ, ε, and ζ Orionis, and bear the Arab names, "Mintaka," the "Belt," "Alnilam," the "String of Pearls" and "Alnitak," the "Girdle," respectively.

δ Orionis is a double star, 23′ of arc south of the celestial equator. ε Orionis is a leading example of stars of the hottest class. Its temperature has been estimated to be 45,000 degrees F.

In astrology these three stars portended good fortune and public honours. Job's name for them was "the Bands of Orion," while the Arabs knew them as "the Golden Nuts," or "the String of Pearls." The fierce Masai

African tribe regarded these stars and those representing the sword of Orion, hanging from the Belt, as "three old widows following up three old men."

The Basuto tribe called the Belt stars "Three Pigs." They have also been known as "the Three Kings," "the Ell," and "the Yard," on account of the line joining them being just three degrees long.

Tennyson thus refers to these stars:

> Those three stars of the airy Giant's zone
> That glitter burnished by the frosty dark.

The Germans designated them "Jacobstaff," "the Staff of St. James," and "the Three Mowers."

The Chinese knew them as "a weighing beam," with the stars in the sword as a weight at one end.

The Greenlanders called them "the Seal Hunters," bewildered when lost at sea, and transferred together to the sky; and to the Eskimos these stars represented the three steps cut in a steep snow bank by some celestial Eskimo to enable him to reach the top.

The early Hindus called these stars "the three-jointed arrow," and the native Australians regarded them as "young men dancing."

In comparatively modern times, 1807, the University of Leipsic, disregarding all ancient appellations, christened these famous stars "Napoleon." An Englishman retaliated by calling them "Nelson," but these names have not been recognised by the world at large, nor do they appear on star maps or globes.

Seamen have called these stars "the Golden Yard Arm."

Tennyson simply referred to them as "the three stars."

In mythology they represent the arrow that despatched Orion. Other names for them are "The Rake," "the Three Marys," and "Our Lady's Wand."

A line drawn through them and prolonged southward passes near the brilliant and famous Sirius.[1]

Three fainter stars in this constellation have also attracted world-wide attention. They form a small triangle and are located in the head of the mighty hunter. The brightest is λ Orionis, a double star. Its Arab name, "Meissa," means "the Head of the Giant." The original name for the star, says Allen, meant "a white spot."

In astrology these three stars were unfortunate in their influence on human affairs. They constituted the Euphratean lunar station known as "the Little Twins," and the Hindu station called "the Head of the Stag."

In China these stars were known as "the Head of the Tiger." Manilius thus refers to them:

> In the vast head immerst in boundless spheres
> Three stars less bright but yet as great he bears,
> But further off remov'd, their splendours lost.

Colas mentions an interesting fact in connection with the triangle formed by these stars, which reveals a current optical delusion. No ordinary observer would imagine that the moon could be contained in this triangle, but such is the fact, for the moon, which to the uninstructed observer appears about the size of "a dinner plate," should be seen as a circle a half-inch in diameter fifty-seven inches away.

σ Orionis is a glorious multiple star, possibly the finest example of its type.

Serviss in his delightful book, *The Pleasures of the Telescope*, thus extols the praises of σ Orionis: "He must be a person of indifferent mind who after looking with unassisted eyes at the modest glimmering of this little star, can see it as the telescope reveals it without a thrill of wonder and a cry of pleasure. The glass, as by a touch of magic,

[1] ζ Orionis is deserving of mention. It is a triple star, the second largest star being of such a peculiar colour as to defy description. Struve called it "ruddy-olive."

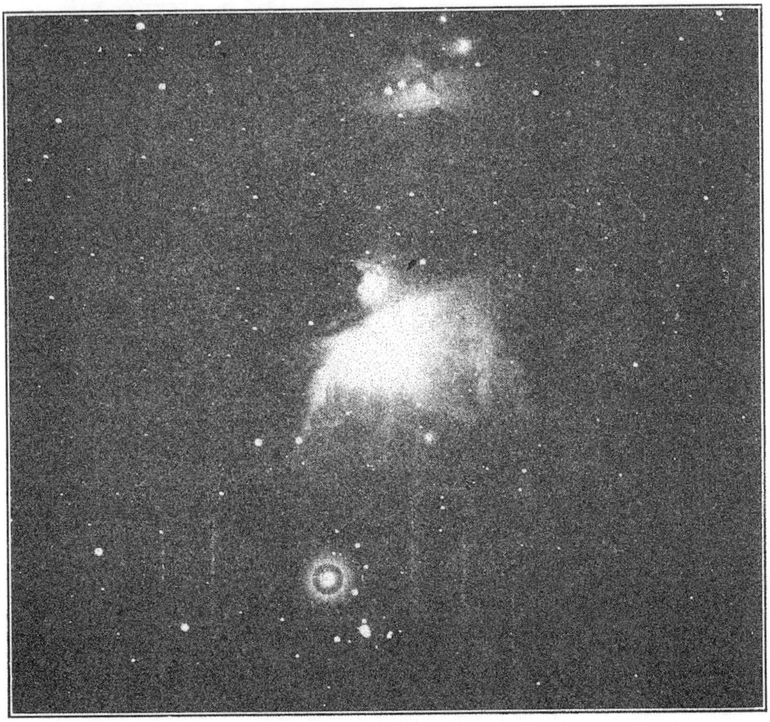

Great Nebula in Orion
Harvard College Observatory

changes it from one into eight or ten stars, and these stars exhibit a variety of beautiful colours charming to behold. However we look at them, there is an appearance of association among these stars, shining with their contrasted colours and their various degrees of brilliance, which is significant of diversity of conditions and circumstances under which the suns and worlds beyond the solar walk exist."

It remains to mention what is probably the most interesting telescopic object, and certainly the most satisfactory to view of its kind in all the heavens, the Great Nebula in Orion. It is situated in the so-called "Sword" of the Giant which hangs pendent from the Belt, and surrounds the star θ Orionis. No description can give an adequate picture of the sight of this wonderful object even in a small telescope. The star θ is divided by the telescope into six stars, four of which can be seen with fairly low power, and compose the well-known "Trapezium."

The nebula itself covers a space equal to the apparent size of the moon, but nebulosity extends over a much greater area. Its spectrum is purely gaseous, and its mass is said to be 4.5 million times that of the sun.

Serviss thus refers to it: "Nowhere else in the heavens is the architecture of a nebula so clearly displayed. It is an unfinished temple whose gigantic dimensions, while exalting the imagination, proclaim the omnipotence of its builder. But though unfinished it is not abandoned. The work of creation is proceeding within its precincts. There are stars apparently completed, shining like gems just dropped from the hand of the polisher, and around them are masses, eddies, currents, and swirls of nebulous matter yet to be condensed, compacted, and constructed into suns. It is an education in the nebular theory of the universe merely to look at this spot with a good telescope. If we do not gaze at it long and wistfully, and return to it many times with unflagging interest, we may be certain that there is not the making of an astronomer in us."

A fitting conclusion to this sketch of Orion and its stars,

is a quotation from Mrs. Martin's *Friendly Stars* respecting the constellation: "With all its wonders and its beauties it is not strange that Orion should be one of the most familiar and most admired of all the constellations. It is in the centre of the Galaxy that marches in brilliant procession across the winter skies. We watch for it between nine and ten o'clock in the evening late in October, and our first view is of the curved line of faint twinkling stars that outline the left arm and the lion's skin.

Then one jewel after another emerges from the storehouse below the horizon until the whole splendid figure is before us. Its arrival is an announcement that the outdoor season is past and that the nights are becoming more and more frosty and that the gorgeous tapestry with which the autumn hills seem covered will soon fade away and give place to the lovely low tones of winter."

Pegasus
The Flying Horse

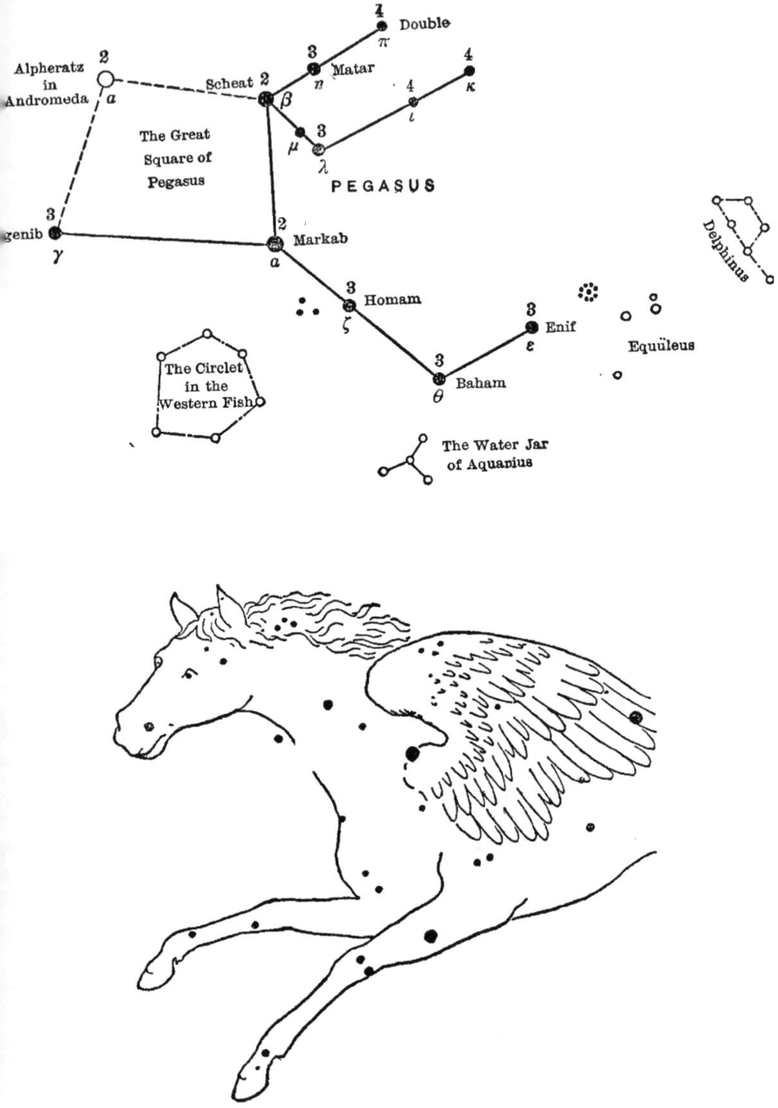
Alpheratz
in
Andromeda
Scheat
Double
Matar
The Great
Square of
Pegasus
PEGASUS
genib
Markab
Homam
Enif
Baham
Equuleus
Delphinus
The Circlet
in the
Western Fish
The Water Jar
of Aquarius

PEGASUS

PEGASUS
THE FLYING HORSE

Then with nostrils wide distended,
  Breaking from his iron chain,
And unfolding far his pinions,
  To those stars he soared again.

LONGFELLOW'S "Pegasus in Pound."

ONLY a part of the figure of a horse appears in this very ancient constellation, and, strangely enough, the horse is always represented reversed, with the forefeet pawing the sky. Pegasus is therefore often referred to as "The Demi-Horse," or "the Half Horse," the steed of the mighty Nimrod.

In mythology this is the celebrated horse that sprang from the blood of the Medusa, which dropped into the ocean after Perseus had severed her head.

According to Hesiod he received his name from his being born near the sources of the ocean, the name being derived from the Greek words πηγαι, meaning the "springs of the ocean," or πηγος, meaning "strong."

Ovid claims Mount Helicon as the home of Pegasus. It was here that, by striking the ground sharply with his hoof, he caused the waters to gush forth, the fabled spring of "Hippocrene."

The poetic steed
With beamy mane, whose hoof struck out from earth
The fount of Hippocrene.[1]

Bryant.

---

[1] Longfellow calls poetic inspiration "a maddening draught of Hippocrene."

Pegasus was tamed by Neptune or Minerva, and was a great favourite with the Muses. He was given to Bellerophon, son of Glaucus, King of Ephyre, to aid him in conquering the Chimæra.

Bellerophon succeeded in destroying the monster, and then attempted to fly up to heaven on his winged steed. Jupiter, angered by his presumption, caused an insect to sting Pegasus, which brought about the fall of his rider. Wordsworth thus mentions the episode:

> Bold Bellerophon (so Jove decreed
> In wrath) fell headlong from the fields of air.

Pegasus, freed of his burden, continued his flight up to heaven, and Jupiter accorded him a place among the constellations.

It is said that Pegasus bears for Jupiter the lightning and thunder.

> Now heav'n his further wand'ring flight confines,
> Where, splendid with his num'rous stars, he shines.
>
> Ovid's *Fasti.*

The fact that Pegasus was especially favoured by the Muses has given rise to the expression often heard, "to mount Pegasus," and every poet it is said must drink of the fountain created by his hoof blow, before he can expect to soar on Pegasean wing. As Spenser says:

> Then whoso will with virtuous wing essay
> To mount to heaven, on Pegasus must ride,
> And with sweet Poet's verse be glorified.

Brown regards Pegasus as the steed of Poseidon, the Charioteer, rising out of "the great deep," or "sea," as this region of the sky was called by the ancients.

Aratos gives us the following description of the Flying Horse:

> He 's not four footed; with no hinder parts
> And shown but half, rises the sacred Horse.

Photo by Anderson

Bellerophon and Pegasus at Rome

They say that he to lofty Helicon
Brought the pure spring of copious Hippocrene.
For upon Helicon no stream flowed down
Till the Horse smote it, there abundant waters
Gushed at the stamp of his fore-hoof. The shepherds
First called it Hippocrene, the Horse's Fountain.
Still from the rock it pours not far from where
The Thespians dwell; thou seest it, but the Horse
Circles in heaven and there thou must behold him.

Plunket thinks there is some support in Grecian and in Vedic legend to be found for the opinion that the original position of Pegasus was upright, and not reversed. Though the Horse appears reversed on the Grecian astronomic sphere, he does not appear so on any artistic monument, vase, or coin.

In the Rig-Veda we read of a swift horse belonging to the Aswins, who, it is said, filled a hundred vases with sweet liquor, an allusion to the fount of Hippocrene.

Max Müller has pointed out that the Aswins possessed a horse called "Pagas," and they are represented by the stars α and β Arietis. If we look at Pegasus in the sky, and observe how closely following that constellation the bright stars that mark the head of Aries appear, we shall easily understand how these Aswins might have, by Vedic bards, been imagined as possessing and driving in front of them the swift steed Pegasus.

As Pegasus is now represented in the heavens, his hoofs do not appear to touch any stellar representation of a fountain or vase, but if the figure is reversed, we find that the forefoot of the Horse would be close to the water-jar of Aquarius, the source of a great stream of water that flows down the steeps of the southern sky.

In the Aswameda hymns in the Rig-Veda, there is a reference to the sacrifice of a horse, and this is thought to refer to the symbolic sacrifice of the winged Horse of the constellation Pegasus.

Plunket thinks the legend of the fount of Hippocrene

dates from 3000 B.C., and the invention of the constellation a thousand years earlier.

In regard to the reversal of the figure, the general opinion is that the figure of the Horse which has come down to us is the original design.

There is a special significance in the star groups that combine two figures in one, or depict merely half a figure. Thus we have the constellation of the Centaur, half man, half horse. This shows that the figure of the horse was familiar to the inventors of the constellations. In Pegasus we find only a half horse; there certainly were plenty of stars and space sufficient to depict the perfect figure, therefore there was some good reason for leaving it out.

Again, we find among the constellation figures half a Bull, only part of a Ship, and a Sea Goat, half fish, half goat. Whatever was the intention in thus depicting these star groups, they certainly furnish additional evidence of a deliberate plan in the minds of the designers and inventors of the star pictures, as everywhere else in the starry skies we find complete figures.

A suggestion has been made, that only half the Horse is shown because the other half is supposed to be obscured by clouds, and that the figure thus depicted conveys a better idea of a horse soaring to the skies. This view is certainly a plausible one.

Allen tells us that Ptolemy mentions the wings of the Horse as well recognised in his day. The winged horse appears to have been a favourite decorative figure, and appears on early Etruscan vases, and on many pieces of pottery found in the valley of the Euphrates. It also appears on coins of Corinth from 500 to 430 B.C. and on a well-known Hittite seal.

The Greeks called the constellation Ἵππος, and in the Alphonsine Tables it was "Alatus," meaning "winged." Apparently at one time the foreleg of Pegasus was considerably extended, as π Cygni bears an Arab name signifying "the hoof of the horse."

Dr. Seiss regards Pegasus as representing the Messenger of Glad Tidings. Jewish legends made it the horse of the mighty Nimrod, and it is also said to represent the ass on which Christ rode in triumph into Jerusalem. Schiller thought this figure represented St. Gabriel.

Bochart claimed that the word Pegasus is a compound of the Phœnician "pag" or "pega," and "sus," meaning the Bridled Horse, used for the figurehead on a ship. It has also been said that Pegasus was of Egyptian origin, from "pag," to cease, and "sus," a vessel, thus symbolising the cessation of navigation at the change of the Nile flow. Here we find Pegasus regarded as the sky emblem of a ship in the very place in the heavens where we should expect to find a craft of some sort, in the part of the sky anciently called "the Sea."

α Pegasi is known as "Markab," an Arab word for a saddle or a ship. It might possibly be that Pegasus, a ship, is the reduplication of Argo, the constellation Ship. In the case of Argo, we find it stranded on a rock. Pegasus is close to the stream pouring from the water-jar of Aquarius, which may represent, as has been supposed, the Flood. It certainly seems more logical to regard Pegasus as a ship rather than half a horse, inasmuch as we have two equine figures in Centaur and Sagittarius, and only one ship, Argo. Reduplication plays such an important part in the constellation designs that we might well expect to find it applicable in the case of the Ship.

An imaginary line connecting α, β, and γ Pegasi and α Andromedæ, a star common to the constellations Andromeda and Pegasus, forms a quadrilateral known as "the Great Square of Pegasus," one of the stellar landmarks.

α Pegasi, or Markab, is one of the so-called lunar stars much observed in navigation. In astrology it portended danger to life from cuts, stabs, and fire. It is on the meridian at 9 P.M. Nov. 3d.

γ Pegasi, called "Algenib," meaning the "wing" or "side," is one of "the Three Guides," the stars that are situated almost on the prime meridian; the other two guides are α Andromedæ and β Cassiopeiæ.

ε, ξ, η, and θ Pegasi bear respectively the following Arab names: "Enif," the nose, "Homam," the lucky star, "Matar," the fortunate rain, "Baham," the good luck of the two beasts.

Most of the faint stars in the constellation have received individual names, an indication of the importance of the constellation in early times.

Within the area of the Great Square Argelander counted about thirty naked eye stars, while Schmidt, observing at Athens, counted one hundred and two.

The writer acknowledges his indebtedness to Prof. W. W. Campbell, of the Lick Observatory, for the following information concerning the stars that form the Great Square of Pegasus:

Alpha Andromedæ is a spectroscopic binary star whose two components revolve around their mutual centre of mass in ninety-seven days, and the system as a whole is approaching the solar system with a speed of 13 km. per second.

Alpha Pegasi is a spectroscopic binary. The observed velocity is variable, and not enough observations have been secured to determine the velocity of the system.[1]

Midway between the head of the Flying Horse and the Dolphin, is a rectangular-shaped figure composed of fourth and fifth magnitude stars, which forms the asterism known as "Equuleus," "the Foal," "the Little Horse," or "the Horse's Head."

The head only of this equine figure is represented, and like the winged Horse appears in an inverted position.

Geminus mentions Equuleus as having been formed by Hipparchus. Ptolemy catalogued it as Ἵππου προτομή. The

[1] Beta Pegasi is receding from the solar system with a velocity of 8 km. per second. Gamma Pegasi is receding from the solar system with a velocity of 5 km. per second.

Arabs called it "Part of a Horse," while with the Hindus it was another of their Aswini, "the Horsemen."

In mythology, according to Allen, Equuleus is said to represent Celeris, the brother of Pegasus, given by Mercury to Castor; or Cyllarus, given to Pollux by Juno, or the creature struck by Neptune's trident from the earth when contesting with Minerva for superiority; but it also was connected with the story of Philyra and Saturn.

Cæsius thought that Equuleus represented the King's Horse that Haman hoped for as told in the book of Esther. It was also thought to represent the mystic Rose.

The asterism comes to the meridian at 9 P.M. on the 24th of September.

δ Equulei is a triple and binary star. The two largest stars form a system noted as the quickest in orbital revolution of all the binaries in the heavens, save two. Its period according to Prof. Hussey is about 5.7 years. The components are so close that they can only be separated by the largest telescopes.

Perseus
The Champion

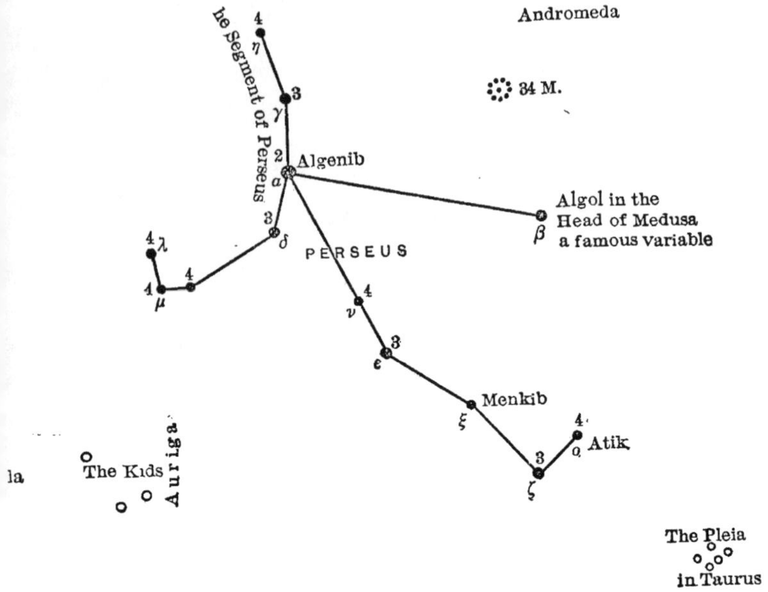
Andromeda
34 M.
he Segment of Perseus
Algenib
Algol in the
Head of Medusa
a famous variable
PERSEUS
Menkib
Atik
la
The Kids
Auriga
The Pleia
in Taurus

PERSEUS

PERSEUS
THE CHAMPION

> Perseus seek for by her feet
> Which ever at his shoulder are revolving.
> Tallest of all his compeers at the North
> He towers. His right hand stretches toward the chair
> Of his bride's mother.
>
> FROTHINGHAM'S *Aratos.*

IN the legend of the sacrifice of Andromeda previously related Perseus figures as the hero, and hence we find his constellation situated close to the groups that represent the unfortunate maiden and the ferocious monster that sought to destroy her.

We can never be sure whether the constellation suggested the legend, or the legend the constellation. It may be, as one authority points out, that the legend was suggested by the fact that the constellation Perseus, rising before Andromeda, seems to deliver it from the night, which might well be depicted as a monster, such as appears in the figure of the constellation Cetus.

It seems, however, as if there were a deeper significance in this story that the stars illustrate, of a conflict between man and beast, with a human sacrifice at stake, and the eventual triumph of man.

We find among the stellar figures many conflicts of this nature depicted, and in no case do we find man overcome in the struggle that he is engaged in.

Hercules and Ophiuchus are seen respectively trampling underfoot or holding securely gigantic reptiles. The Archer is about to slay the Scorpion, and Orion threatens undismayed the advancing Bull. Perseus, flying from his victory over the dread Medusa, slays the monster of the deep, and rescues the fair Andromeda.

This universal victory of humanity over the animal world, depicted in the constellations, is one of the chief features of these time-honoured configurations, and is clearly indicative of the fact that the ancient star groups are the product of design and not chance.

Perseus, because of his gallantry, is known as "the Knight Errant of Mythology." The hero was the son of Jupiter and Danaë, and the favourite of the gods. His successful encounter with the Medusa rendered his name immortal, and at his death, it is said, he was transported to the starry skies, where he appears with upraised sword, holding the severed head of the Medusa up to the gaze of all mankind, for all time.

Plunket is of the opinion that the constellation was invented about 1433 B.C., as at that time the star Algol, the well-known variable and most interesting star in the constellation, exactly marked the equinoctial meridian.

"The northern latitude 40° N., suitable for the imagining of this constellation, and the name 'Perseus' seem to point," says Plunket, "to an Iranian school of astronomers as the probable originators of this figure."

Aratos, in an allusion to Perseus, describes him as "stirring up the dust in heaven," so great was his haste to rescue Andromeda. It has been suggested that the dust may be represented by the Milky Way, in a part of which the constellation is located.

Without doubt the story of Perseus was well known in Greece anterior to the 5th century B.C., for Euripides and Sophocles each wrote a drama based on Andromeda's history.

The Arabs called the constellation "Bearer of the Demon's Head," which is represented by the star β Persei, known generally as "Algol." Perseus has also been called "the Rescuer," and "the Destroyer." Dr. Seiss regards the figure as symbolising the Redeemer of Mankind, and others

Photo by Hanfstaengl

Perseus and Andromeda

Painting by Rubens

have claimed that Perseus represents David with the head of Goliath, and the Apostle Paul with his sword and book.

The constellation is 28° in length, one of the most extended in the heavens. Its principal stars form a curved line that bears the name of "the Segment of Perseus," a figure that is almost as much of a stellar landmark as the Great Square of Pegasus.

The most interesting star by far that the constellation contains is the variable "Algol," the so-called "Demon Star," or "Blinking Demon."

The variability of this remarkable star was first scientifically noted by Montanari in 1670, but it is tolerably clear that these light variations had been detected long before his day. Indeed the winking of this star, so to speak, probably influenced those who christened it, so that they likened it to the eye of some great demon peering down through space seeking his prey.

Goodricke, in 1782, was the first one to advance the eclipse theory to account for the variations in the light of this star, and since that date it has been under the constant observation of trained observers.

In 1880 Pickering reaffirmed the eclipse theory, and Vogel subsequently proved the theory unquestionably correct by means of that wonderful instrument the spectroscope.

During 2.5 days Algol is constant at magnitude 2.3. It then begins to diminish in brilliance, at first gradually, and afterwards with increasing rapidity to 3.5 magnitude during a period of about nine hours; its total period being stated as two days, twenty hours, forty-eight minutes, and fifty-five seconds.

In accordance with the eclipse theory to account for the variations of light in Algol, it has been proved that this star is accompanied by a great dark satellite as large as our sun, which at regular intervals passes between us and Algol, cutting off a portion of its light. Algol is said to be one million miles in diameter, while the diameter of the satellite is given as eight hundred thousand miles. The distance between these enormous bodies is three million miles. Both stars are probably surrounded by extensive atmospheres and their united mass is estimated to be two-thirds that of our own sun.

It is said that the famous astronomer Lalande, who died in Pàris in 1807, was wont to remain whole nights, in his old age, upon the Pont Neuf, to exhibit to the curious the variations in the brilliancy of the star Algol.

In astrology Algol was considered the most unfortunate and dangerous star in the heavens.

Among the Hebrews Algol was said to represent Adam's mysterious first wife, Lilith. The star has also been called "the Medusa's or Gorgon's Head," "Satan's Head," "the Spectre's Head," "Double Eye," and the Chinese knew Algol by the unsavoury name of "Piled up Corpses." The Arabs called the star "Al-Ghul," the Demon or "Fiend of the Woods," from which our word ghoul is derived.

Algol is a Sirian star, and is approaching our system at the rate of a mile a second. It is on the meridian at 9 P.M. Dec. 23d, and at that time is almost exactly in the zenith of New York City.

There are two beautiful star clusters in this constellation, situated in the "Sword Hand of the Champion," about midway between the "Segment of Perseus" and Cassiopeia.

Hipparchus refers to them as "a cloudy spot," while Ptolemy called them "a dense mass." They are visible to the naked eye, and present a beautiful appearance in an opera-glass. Seen through a telescope the glory of the sight is indescribable. In one of these clusters at least one hundred stars may be seen in an area less than one quarter as broad as the face of the full moon.

The well-known Perseid meteor shower, with its maximum about August 10th, radiates from this constellation. These meteors are sometimes called "the Tears of St. Lawrence," and the shower has been recorded as far back as the year 811.

α Persei, a star of the second magnitude, bears the name "Algenib," meaning the "Side." It has also been called

Photo by Brogi

Perseus and Medusa
Bronze by Cellini at Florence

"Marfac," meaning the "Elbow." It is flanked on either side by a bright star, and in this respect resembles Altair, the first magnitude star in Aquila.

ε Persei is a double star, "especially interesting," says Serviss, "on account of an alleged change of colour from blue to red which the smaller star undergoes coincidently with a variation of brightness."

η Persei is also a double star, noteworthy as having three faint stars on one side nearly in a line and one on the other forming a miniature representation of Jupiter and his satellites.

ξ and ο Persei bear respectively the names "Menkib" and "Atik."

"The Milky Way around Perseus," says Burritt, "is very vivid, being undoubtedly a rich stratum of fixed stars, presenting the most wonderful and sublime phenomenon of the Creator's power and greatness." Kohler, the astronomer, observed a beautiful nebula near the face of Perseus, besides eight other nebulous clusters in different parts of the constellation.

About midway between β and δ Persei there appeared Feb. 21, 1901, a *nova* or new star. It was discovered by Dr. Anderson of Edinburgh and when first seen was of 2.5 magnitude. It shone with a bluish-white light and two days after its discovery it was brighter than Capella, having in three days increased its brightness 25,000 fold. All *novæ* are temporary and rapidly diminish in brightness. Following the usual course Nova Persei became invisible to the naked eye in six weeks and its spectrum soon became nebulous. So terrific was the heat evolved that the gases expanded outward with a velocity of over 2000 miles a second, and the distance was so great that its light only reached us after a period estimated at 300 years, hence the collision which we witnessed by the advent of this new star must have occurred about the year 1600.

Nova Persei was the most brilliant star that has appeared since 1604.

20

Pisces
The Fishes

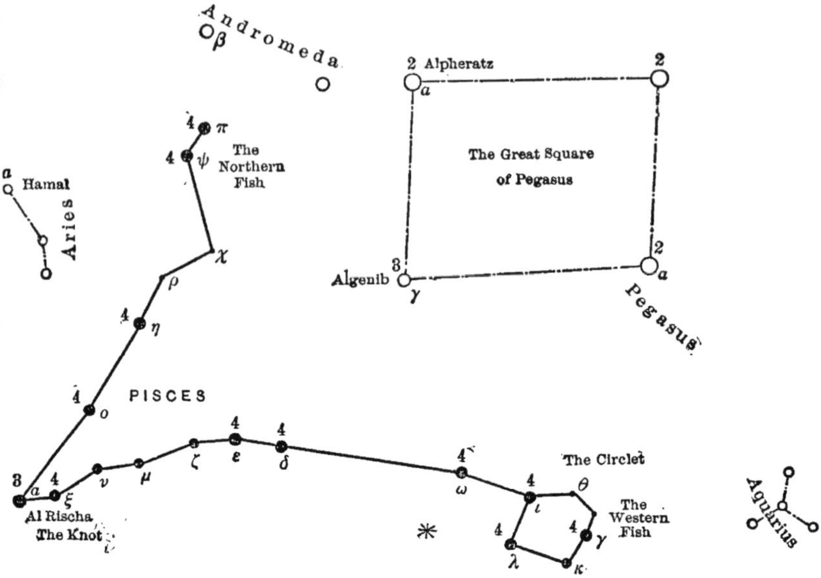
Andromeda
2 Alpheratz
The Great Square of Pegasus
Algenib
Pegasus
Hamal
Aries
The Northern Fish
PISCES
The Circlet
The Western Fish
Al Rischa
The Knot
Aquarius

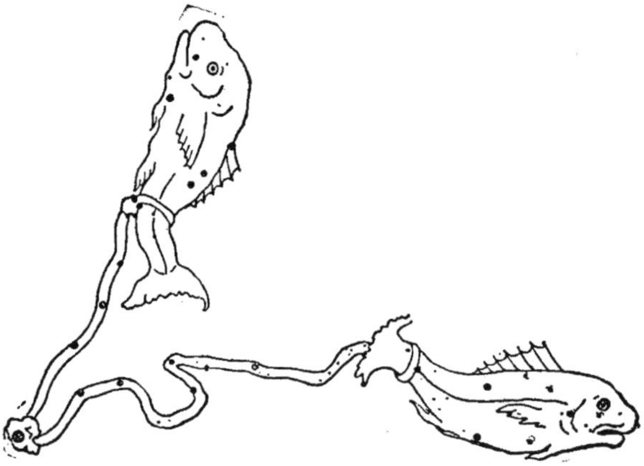

PISCES

PISCES
THE FISHES

> The Fishes shine one higher than the other,
> From each of them extends as 't were a band
> That fastens tail to tail, as wide it floats,
> And one star large and brilliant clasps its ends,
> The Heavenly Knot 't is called.
>
> FROTHINGHAM'S *Aratos.*

OWING to the Precession of the Equinoxes, the constellation of the Fishes is now the Leader of the Celestial Hosts. The vernal equinox, or the point where the sun crosses the equator in the spring, is situated in Pisces and this point is often referred to as "the Greenwich of the Sky." From it the Right Ascension of all the stars is reckoned.

Pisces is usually represented on the star maps by the figures of two fishes a considerable distance apart; around the tail of each is tied a ribbon, and the ends of these bonds are joined together and tied in a knot, which the star "Al-Rischa," or α Piscium, represents.

According to Greek mythology, Venus and her son Cupid were strolling along the banks of the Euphrates River. They were alarmed at the sudden appearance of Typhon, a terrible giant, whose chief occupation seems to have been to frighten people. To escape the monster, Venus and Cupid leaped into the river and assumed the form of two fishes. To commemorate this event Minerva placed two Fishes among the stars.

In accordance with this myth the constellation was popularly known as "Venus and Cupid."

This legend of the escape of Venus and Cupid from the dread Typhon is analogous to the myth concerning the constellation Capricornus, where, as we have seen, Pan or Bacchus escaped from Typhon by jumping into the river Nile, and assuming the form of a Goat-Fish.

It is comforting to know that Typhon was finally disposed of by the father of the gods, and, according to the myth, he lies crushed to death beneath Mount Ætna.

The Babylonians, Syrians, Persians, Turks, and Greeks all regarded this star group as representing two Fishes, and we find them appropriately placed in the part of the sky known to the ancients as "the Sea," near the Whale, the Dolphin, and the Southern Fish.

Sayce is of the opinion that the dual form of this constellation is due to the double month inserted every six years into the Babylonian calendar.

The two Fishes are known as "the Northern Fish," which lies just south of Andromeda, and "the Western Fish," situated below Pegasus. The former was known to the Chaldeans as "the Tunny," and it is said that there was an important tunny fishery at Cyzicus, which might have influenced the choice of these symbols.

According to the Egyptians this sign denoted the approach of spring and the season for fishing. It is also claimed that the name of the Fishes was derived from the fact that, at the time when the sun entered Pisces, fishes were considered as fattest and most in season for use.

Brown claims that Pisces is a reduplication of the nocturnal sun, the fish sun concealed in the waters. The archaic myth is that of the resumption of the cultivation of the earth after the catastrophe of the Flood.

The Arabs knew the Western Fish as "Al-Hut," the Fish, and they considered the stars in the Northern Fish as part of the constellation Andromeda.

Allen tells us that the Chaldeans imagined the Northern Fish with the head of a swallow. The association of a bird with this constellation is very curious. Among the Peru-

Photo by Hanfstaengl

Venus and Cu id

vians the month of Pisces was represented by two star groups, one called "the Terrace of the Granaries" or "the Doves," a name also given to the Pleiades. This group was figured as a kind of net with numerous meshes. For some unexplained reason the Pleiades seem to have been associated with this sign in the Orient. The other Peruvian asterism was called "Pichu," the Knot, by which name the month was also known, and it was represented by a net enclosing fishes. The connection between Pisces and the Pleiades is emphasised by the analogy in the idea of snaring as applied to both birds and fishes, and Tennyson, though probably unaware of it, expresses the idea in his reference to the Pleiades, when he likens them to "fireflies tangled in a silver braid."

In the Hebrew zodiac Pisces represented the tribe of Simeon, and the Fishes were considered the national constellation of the Jews, as well as a tribal symbol.

Dr. Seiss considers that the Fishes symbolise "the Twofoldness of the Church," while Schiller thought the figure represented St. Matthias.

In astrology Pisces is the House of Jupiter and the Exaltation of Venus. Those born from Feb. 19th to March 20th are its natives. They are supposed to be short, thickset, pale, and round shouldered, with characters phlegmatic and effeminate.

It governs the feet and reigns over Portugal, Spain, Egypt, Normandy, Calabria, etc.

It is a feminine sign and unfortunate. "No sign," says Burritt, "appears to have been considered of more malignant influence than Pisces. The astrological calendar describes the emblems of this constellation as indicative of violence and death. Both the Syrians and Egyptians abstained from eating fish, out of dread and abhorrence, and when the latter would represent anything as odious or express hatred by hieroglyphics, they painted a fish."

The 26th Hindu lunar station lay in this sign, and

contrary to the malignant influence ascribed to the constellation it was designated "abundant or wealthy."[1]

The flower ascribed to Pisces is the daffodil, and the gem the white chrysolite.

The symbol of the sign, ♓, is thought to represent the two fishes joined together. A fish was always the symbol of the early Christian faith, and the figure appears in many of the stained glass windows in the churches of to-day.

When each sign of the zodiac was assigned to one of the twelve Apostles, the Fishes were said to represent St. Matthias.

The Western Fish is represented by a lozenge-shaped figure traced by faint stars, which is known as "the Circlet."

Three distinct conjunctions of Jupiter and Saturn were recorded as taking place in Pisces in the year 747 of Rome. This was the year in which for a long time Christ was supposed to have been born. The claim has been made that the star of Bethlehem was so to speak a composite star, a conjunction in Pisces of the planets Saturn, Jupiter, and Mars.[2]

Jupiter, Saturn, and Venus were all located here in February, 1881. Stöffler predicted in 1524, when these three planets were in conjunction in Pisces, that there would be another Deluge. The season was unusually dry. It was in this constellation that Harding discovered Juno in September, 1804.

The principal star in the constellation is α Piscium, known as "Al-Rischa," meaning the Cord, or "Okda," the

[1] The star ζ Piscium marks the initial point of the fixed Hindu sphere from which longitude was reckoned. This point coincided with the vernal equinox A.D. 570. This date, says Burgess, fixes approximately the commencement of the history of modern Hindu astronomy.

[2] In support of this theory a recent writer has said that "such a conjunction would at once have been interpreted by the Chaldæan observers as indicating the approach of some memorable event, and since it occurred in the constellation of Pisces, which was supposed by astrologers to be immediately connected with the fortunes of Judæa, it would naturally turn their thoughts in that direction."

Knot of the two threads. It marks the knot formed by the joining together of the ends of the ribbons that hold the Fishes fast by the tail. The Arabs knew these two cords as "the Flaxen Thread."

It is a double star which culminates at 9 P.M. Dec. 7th. The remaining stars in the constellation are unimportant.

On a clear night, when the moon is absent from the sky, the lines of stars representing the ribbons can be clearly seen. Starting from the Knot Star, the stars diverge to the east and west, forming a "V"-shaped cleft, into which the Great Square of Pegasus seems about to fall.

Sagittarius
The Archer

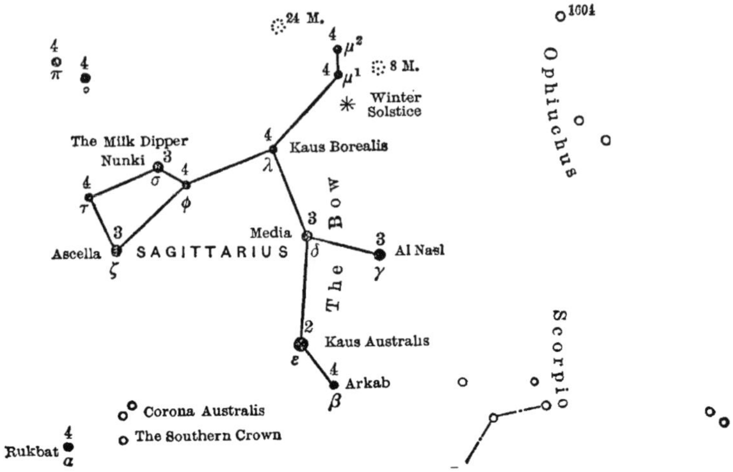

SAGITTARIUS

SAGITTARIUS
THE ARCHER

Midst golden stars he stands refulgent now
And thrusts the Scorpion with his bended bow

OVID.

THE antiquity of this constellation is attested by the fact that it is depicted on ancient Babylonian monuments, and upon the early zodiacs of Egypt and India.

Sagittarius, according to Greek mythology, represents the famous centaur Chiron, son of Philyra and Saturn, who changed himself into a horse to elude his jealous wife, Rhea. Ovid tells us that Chiron was slain by Hercules with a poisoned arrow. Chiron, realising that the wound was incurable, begged Jupiter to deprive him of immortality. The father of the gods granted his request, and placed him among the constellations.

Another legend relates that Apollo urged the moon goddess Artemis to aim a shaft from her bow at a gleaming point on the horizon, which concealed Orion, the mighty hunter. Orion was thus unwittingly slain by Artemis. The constellation Orion is exactly in opposition to the so-called "Bow stars" of Sagittarius, which accounts for this myth connecting the two constellations.

The legend is clearly astronomical in its significance, for in the variant form here depicted, Artemis is represented as sending a scorpion to sting Orion to death, and we find the stars marking the scorpion's sting in very close proximity to the Bow stars of Sagittarius.

On ancient obelisks the figure of an arrow is sometimes seen, which is supposed to be a hieroglyphical representation of this sign. In the Indian zodiac the name of the constellation simply means "arrow."

It is thought that the Egyptians made use of the figure of Hercules to represent this constellation, and that the Greeks chose to substitute the centaur in compliment to Chiron.

The centaur Chiron was sometimes called "the Bull Killer." The astronomical significance of this appellation is as follows: When the constellation Sagittarius rises in the east, it always seems to drive below the western horizon the last stars in the constellation Taurus, the Bull. Thus the Archer, metaphorically speaking, slays the Bull.

Chiron was reputed to be a wonderful archer, and taught the Grecian youths the use of the bow and arrow. He is always represented with bow drawn, aiming a shaft at the heart of the Scorpion. As Manilius puts it:

> . . . glorious in his Cretian bow,
> Centaur follows with an aiming eye,
> His bow full drawn and ready to let fly.

According to Brown this constellation is a solar variant, and we have represented here the rising sun shooting out his shafts across the morning skies. From a fragment of the Euphratean Planisphere it is indicated that the Akkadian name for Sagittarius signified "Light of the White Face," or "Smiting Sun Face." Cuneiform inscriptions designate Sagittarius as "the Strong One," "the Giant of War," and "the Illuminator of the Great City."

There are few constellations in which the figures of the monuments and the descriptions in the tablets show a closer connection between Euphratean and classical forms than in the case of Sagittarius.

The constellation is identified with the Assyrian god Assur and the Median god Ahura. Ahura is generally represented as holding in his hand a ring or crown, and Assur in some examples is also thus depicted. Close to the hand of the Archer we see the ancient Ptolemaic constellation "Corona Australis," the Southern Crown, which is generally represented as a ring-shaped wreath. This accounts for the substitution of the crown or wreath for the bow and arrow.

From approximately 6000 to 4000 B.C. Sagittarius was the constellation in which the autumnal equinoctial point was located, the equinoctial colure passing through the constellations Sagittarius and Taurus. In accordance with this, we find on one of the ancient Assyrian standards the figure of an archer above that of a galloping bull.

Plunket claims that originally only the bow and arrow of Sagittarius were represented for this division of the ecliptic. The first recorded classic figuring of the Archer was in Eratosthenes' description of it as a Satyr. Afterwards it was changed to a Centaur or Bull Killer. The centaurs were an ancient race inhabiting Mt. Pelion in Thessaly.

Longfellow in his "Poet's Calendar" thus refers to the Archer:

> With sounding hoofs across the earth I fly,
> A steed Thessalian with a human face.

The stars ζ, τ, σ, φ, and λ Sagittarii form a figure known as "the Milk Dipper." The Dipper appears inverted and the title is appropriate as it is situated in the Milky Way. This figure was known to the ancients as "the Ladle," and these stars were the objects of special worship in China for at least a thousand years before our era. The Chinese called this figure "the Temple," and Sagittarius was known to them as "the Tiger." The Milk Dipper is also called "the Hobby Horse of Sagittarius."

λ, δ, and ε Sagittarii form the bow of the Archer. This bow has metaphorically been regarded as "the Bow of Promise Set in the Cloud," succeeding the Deluge, the "Cloud" being represented by the Milky Way.

The Arabs called this constellation "the Bow." They imagined the stars in the group represented ostriches passing to and from the celestial river, the Milky Way. The star λ represented their keeper.

It is indeed strange, as Ideler points out, that these non-drinking animals should be found here so close to a river, but the suggestion has been made that these stars represented pasturing cattle, that being the translation of Na'aîm, the title of the 20th lunar station of the Arabs located here.

Some authorities, who claim to explain the origin of the constellations, assert that Sagittarius was so called because, at the time the sun entered it, the hunting season opened, and that this is the Archer or Huntsman. Sagittarius has always been considered the patron of the hunter and the chase.

There is in the figure further evidence of design on the part of the inventors of the constellations, for we find the Horse of Pegasus endowed with wings, which are denied Centaurus and the Archer.

The Jews regarded the Archer as the tribal symbol of Ephraim and Manasseh, while Dr. Seiss calls Sagittarius "a pictorial prophecy of our Blessed Lord." The Archer appears on a coin of Gallienus of about A.D. 260, and Schiller thought the figure represented St. Matthew.

Astrologically speaking, Sagittarius is the House and Joy of Jupiter. Its natives, those born between the dates Nov. 22d and Dec. 21st, are said to be well formed, with fine clear eyes, chestnut hair, and oval fleshy face. They are generally of a jovial disposition, active, intrepid, generous, and obliging. It governs the legs and thighs, and reigns over Arabia, Spain, Hungary, Moravia, Cologne, etc. It is a masculine sign and fortunate. The goldenrod is the flower, and the carbuncle is the significant gem.

Ampelius associated it with the south wind, and the colour yellow was attributed to it.

α Sagittarii bears the name "Rukbat," meaning the "Archer's Knee."

β, a double star, was designated "Arkab," the "Tendon,"

and α and β were known to Kazwini as "two desert birds."

γ was called "Al-Nasl," meaning the "Point," *i. e.*, of the arrow which the Archer aims at the Scorpion.

δ and ε are double stars. The former was known as "Kaus Meridionalis" or "Media," meaning the "Middle," *i. e.*, of the bow. The latter was "Kaus Australis," the southern (part of the) bow.

ζ was called "Ascella," the "Armpit."

λ bore the title "Kaus Borealis," meaning the northern (part of the) bow.

μ^1, a triple star, and μ^2 mark the point of the winter solstice.

σ, known as "Nunki," also bore the title, "the Star of the Proclamation of the Sea."

The symbol of the sign Sagittarius (♐) indicates the arrow and part of the bow.

ω and three stars near it form a small quadrangle on the hind quarters of the horse, which bears the name "Tere-Bellum." This figure was known to the Chinese as "the Dog's Country."

There are several fine naked-eye star-clusters in this constellation, which also contains the celebrated "Trifid Nebula," discovered in 1764.

An exceedingly brilliant *nova* is said to have appeared low down in Sagittarius in the year 1011 or 1012, which was visible for three months. The appearance of this star was recorded in the astronomical records of China.

Sagittarius also contains one of the so-called "Coal Sacks" in the Milky Way, dark spots where no stars appear. One of these is near the stars γ and δ Sagittarii, showing but one faint telescopic star.

There is another remarkable spot near the star λ Sagittarii, of which Prof. Barnard writes: "It is a small black hole in the sky. It is round and sharply defined. Its measured diameter on the negative is 2.6′. On account of its sharpness and smallness and its isolation, this is perhaps the most remarkable of all the black holes with which I am acquainted. It lies in an ordinary part of the Milky Way, and is not due to the presence or absence of stars, but seems really to be a marking on the sky."

Scorpio
The Scorpion

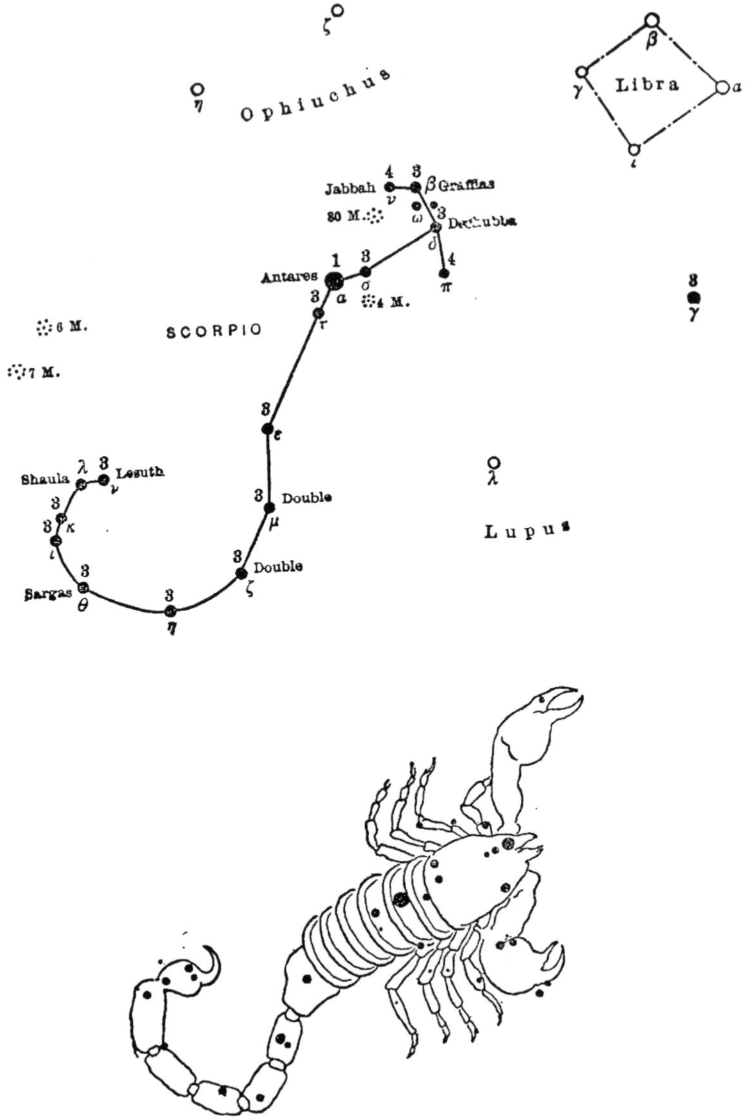

Ophiuchus
Libra
Jabbah
Graffias
80 M.
Antares
4 M.
SCORPIO
6 M.
7 M.
Shaula
Lesuth
Double
Double
Sargas
Lupus

SCORPIO

SCORPIO
THE SCORPION

> There is a place above where Scorpio bent,
> In tail and arms surrounds a vast extent.
> In a wide circuit of the heavens he shines,
> And fills the place of two celestial signs.
>
> OVID.

THIS is the famous Scorpion which sprang out of the earth at the command of Juno, and stung Orion, the mighty hunter, of which wound he died.

It has been suggested that the inventors of the constellations might have placed the Scorpion in this region of the zodiac to denote that when the sun enters this sign, the diseases incident to the fruit season would prevail, since autumn, which abounded in fruit, often brought with it a great variety of diseases, and might be thus fitly represented by that venomous creature the Scorpion, who, as he recedes, wounds with a sting in his tail. However, there seems a deeper significance for the name and position of this constellation, which Maunder points out. At midnight at the spring equinox, the Scorpion was for the ancients who designed the star groups on the meridian in the south, and the Dragon was in like manner on the meridian in the north, so they provided another hero, the Serpent Holder, to trample down the Scorpion in the south, just as Hercules treads on the Serpent in the north, the heads of the two heroes being represented by stars in the zenith. Both the unknown warriors therefore were pictured in those primitive ideas as erect, but for many generations Hercules has appeared to us hanging downwards in the sky.

There can be little doubt that these four figures are

connected, and they are so arranged that whichever way we view the heavens, facing the meridian, we see a giant treading on a serpent.

The Serpent-Bearer presses down the head of the Scorpion at the point where the colure, the equator, and the ecliptic intersected. This is significant, and the arrangement of these constellations was unquestionably the result of a deliberate plan.

For some reason the equator, the colures, the zenith, and the Poles were all marked out by serpentine or draconic forms. In this case the Scorpion is clearly depicted as curling his sting upwards to wound the giant's heel. We see again a seeming illustration of the Biblical utterances: "I will put enmity between thee and the woman and between thy seed and her seed. It shall bruise thy head and thou shalt bruise his heel."

Scorpio is one of the most ancient of the constellations, originally much extended in the direction of Virgo, the claws of the Scorpion occupying the region of the sky where we now see the constellation Libra. In early times this sign was represented by various symbols, sometimes by a snake or crocodile, but most commonly as a Scorpion.

Brown tells us that the Scorpion, like the Crab, was originally a symbol of darkness, and the original strife between the Orion-Sun and the Scorpion Darkness. This is astronomically reduplicated in the constellations Orion and Scorpio, where the stars in the former group appear to be routed by the rising stars of the Scorpion. This symbolism seems to be the foundation for the Greek legends concerning the death of Orion occasioned by the Scorpion's sting.

As the Scorpion rises in the eastern sky, Orion, as if in fear, disappears from view in the west. The Scorpion had much to answer for, as, besides slaying the mighty hunter, he is said to have stung the horses Phaëton drove on his disastrous ride in the chariot of the sun.

On the early Euphratean monuments is found the

figure of a lamp, below which and almost touching it appears a scorpion with large claws. The stars in the claws form a circular figure, and some authorities claim they represent the waning sun.

Aratos speaks of "the fiery sting of the huge portent Scorpio in the south wind's bosom."

Sir Wm. Drummond asserted that in the zodiac which the patriarch Abraham knew, Scorpio was the Eagle. There is a claim made, and it seems not improbable, that the figure of the Cherubim in its fourfold character appears in the constellations. It was described by Ezekiel as the likeness of four living creatures, the lion, the calf, the third creature having the face as of a man, the fourth like a flying eagle. It is certainly significant and can hardly be a coincidence that we find such figures in the four most important positions in the sky. The constellations were originally so designed that the sun at the time of the summer solstice was in the middle of the constellation Leo, at the time of the spring equinox in the middle of Taurus, and at the time of the winter solstice in the middle of Aquarius. The fourth point, that held by the sun at the autumnal equinox, would appear to have been already assigned to the foot of the Serpent-Bearer as he crushes down the serpent's head. Here we find the Scorpion, the very constellation that, according to Drummond, Abraham knew as the Eagle.

Some authorities claim that Aquila is the flying eagle, the semblance of the fourth face of the Cherubim, but in view of the fact that Antares, the brilliant first magnitude star in the Scorpion, is always known as one of the so-called four "Royal Stars," known as such from remote antiquity, there seems to be some ground for the argument that the constellation Scorpio was originally considered to represent the eagle.

Allen tells us that the Akkadians called this constellation "the Seizer" or "Stinger," and "the Place where One Bows Down." The Arabs, Persians, and Turks all regarded it as a Scorpion, while by the Mayas, an ancient race residing in Yucatan, it was regarded as "the sign of the Death god."

In the Hebrew zodiac, Scorpio is referred to the tribe of Dan, because it is written, "Dan shall be a serpent by the way, an adder in the path." The Egyptians fixed the entrance of the sun into Scorpio as the commencement of the reign of Typhon, and when the sun was in this sign the death of Osiris was lamented. Some commentators have located in this constellation the Biblical "Chambers of the South."

The Scorpion is clearly indicated on the celebrated zodiac of Denderah, and the constellation has been likened to a cardinal's hat, and a kite. It certainly bears a striking resemblance to the latter.

Early Christians claimed that this figure represented the Apostle Bartholomew.

Scorpio was known to astrologers as "the accursed constellation," the baleful source of war and discord. Like Aries it is the House of Mars and also his joy. Its natives, those born between the dates Oct. 23d and Nov. 22d are said to be strong, corpulent, and robust, with large bones, dark curly hair, dark eyes, middle stature, dusky complexions. They are usually reserved in speech. It governs the region of the groin, and reigns over Judæa, Norway, Barbary, Morocco, Messina, etc. It is a feminine sign and unfortunate. The red carnation is the flower and the topaz the gem.

The weather-wise thought that this constellation exerted a malignant influence, and was accompanied by storms, but the alchemists held Scorpio in high regard, for only when the sun was in this sign could the transmutation of iron into gold be performed.

Scorpio is in a region of the heavens famous for the appearance of *novæ*, the wonderful temporary stars that occasionally flash upon our view the light that spells a great conflagration or mighty cataclysm far out in space, the enormity of which is beyond our comprehension.

α Scorpii is known as "Antares," from the Greek words

ἀντι Ἄρης, meaning "similar to" or "a rival of Mars," doubtless in reference to its reddish hue. The ancient Hindus called the star "the ruddy."

The Arabs knew it as "the Scorpion's Heart," and even now it is often called "Cor Scorpii," the heart of the Scorpion.

> The heart with lustre of amazing force
> Refulgent vibrates; faint the other parts,
> And ill-defined by stars of meaner note.

Antares was one of the four "Royal Stars" of Persia, 3000 B.C. Chinese documents of great antiquity refer to Antares as "the Fire Star" or "Great Fire." It was also known as "the red or unlucky star." In central Asia it was known as "the Grave Digger of Caravans," because as long as the caravans observed its rising with Orion in the morning, robbers and death followed the stations.

Some of the ancient temples of Egypt were oriented to Antares, edifices that were built thousands of years before the Christian era, and Greek temples at Athens, Corinth, Delphi, and Ægina contain architectural features of a like nature, showing clearly that the star Antares played an important part in the temple worship.

On the Euphrates Antares was known as "the Lord of the Seed," "the Lusty King," "the Vermilion Star," and "the Day-Heaven-Bird." This latter title seems to confirm the idea that this constellation was originally intended to represent an eagle.

Jensen claims that Antares is the "Lance Star" referred to in the 38th chapter of the book of Job.

Mrs. Martin thus refers to the rising of this ruddy-hued sun: "Before one has really seen the star he becomes conscious of a ruddy glow low in the south-east that at once fastens the attention. It is the face of Antares whose red light shining through the heavy atmosphere is so diffused that it gives a rosy effect to the sky for a considerable distance around the star, like a miniature presentment of the sun as it rises on a hazy morning."

Antares belongs to Secchi's third spectroscopic type of stars, the suns that are slowly growing cold as their fires burn low. Like huge embers they still glow with latent heat, like sullen demons doomed to death these flame-scourged suns await the frigid touch that time bestows on life, be it on this mundane sphere or in the uttermost parts of the firmament.

Antares has a tiny emerald green companion which can be seen in a five-inch telescope. Serviss, in his *Pleasures of the Telescope,* thus refers to it: "Antares carries concealed in its rays a green jewel which to the eye of the enthusiast in telescopic recreation appears more beautiful and inviting each time he penetrates to its hiding-place. . . . When the air is steady and the companion can be well viewed, there is no finer sight among the double stars. The contrast in colours is beautifully distinct—fire-red and bright green. The little green star has been seen emerging from behind the moon ahead of its ruddy companion."

Two or three degrees north of Antares is the location of the discovery of Coddington's Comet C of 1898, the third comet to be discovered photographically.

Antares rises at sunset on the 1st day of June, and culminates at 9 P.M. on July 11th.

The triple star β Scorpii is known as "Graffias," of unknown derivation. Allen points out that the Greek word Γραψαῖος signifies crab, and that the words for crab and scorpion were almost interchangeable in the early days. This may possibly explain the origin of the title of this star. Timochares, it is said, observed an occultation of β by the moon in the year 295 B.C.

The three stars in a line, β, δ, and π Scorpii, seem to have attracted attention in all ages, much as the three stars in Orion's Belt are always associated together. The Hindus figured these stars as a Row or Ridge, and on the Euphrates this group represented the Tree of the Garden of Light, associated with the idea of the Tree of Life in the midst of the Garden of Eden, which has a special significance when it is recalled that Scorpio may be considered as representing one phase of the Cherubim which was set in the Garden of Eden.

λ and υ Scorpii are situated in the sting of the Scorpion, which appears to be raised and about to strike the heel of the Serpent-Bearer. The former was known as "Shaula," meaning the "Sting." This star was regarded as unlucky by the astrologers. υ Scorpii was called "Lesuth." These stars were known as "the two releasers," their rising being supposed to bring relief to those suffering from lingering disease.

The row of stars from μ to υ Scorpii was known to the Polynesian islanders as "the Fishhook of Mani," with which that god drew up from the depths the great island Tongareva. They also regard μ^1 and μ^2 as brother and sister, fleeing from home to the sky when ill treated by their parents, the stars λ and υ, who followed them, and are still in pursuit.

The Chinese knew λ and υ Scorpii as "the parts of a lock." Above these stars are two very fine star clusters visible to the naked eye, and beautiful objects even in an opera-glass. One of these was a great favourite with Sir Wm. Herschel, who discovered that it was a star cluster and not a circular nebula, as Messier had claimed. Herschel considered this cluster the richest mass of stars in the firmament.

Taurus
The Bull

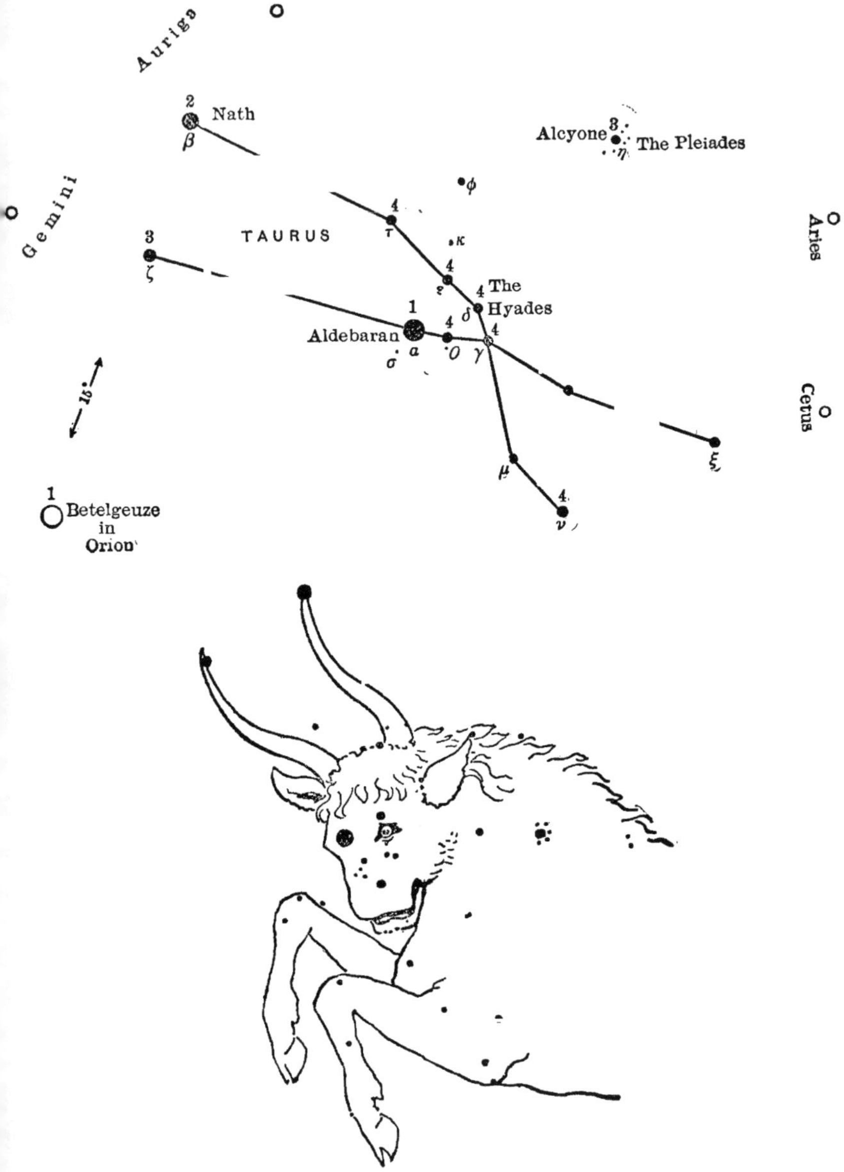

Auriga
Nath
Alcyone
The Pleiades
Gemini
TAURUS
Aries
The Hyades
Aldebaran
Cetus
15°
Betelgeuze
in
Orion

TAURUS

TAURUS
THE BULL

I mark, stern Taurus, through the twilight grey,
The glinting of thy horn,
And sullen front, uprising large and dim,
Bent to the starry Hunter's sword at bay.

TAYLOR.

THERE is every reason to suppose that the constellation Taurus was one of the first to be invented. In ancient Akkadia it was known as "the Bull of Light," and before the time of Abraham, or over four thousand years ago, the Bull marked the vernal equinox. For the space of two thousand years therefore, Taurus was the prince and leader of the celestial hosts.

The sun in Taurus was deified under the symbol of a bull and worshipped in that form, and evidence of this idolatry is seen in the sacred figures found among the ruins of Egypt and Assyria, in the form of a bull with a human face, or a human shape with the face and horns of a bull.

On the walls of a sepulchre excavated at Thebes, Taurus is shown as the first of the zodiacal signs, and the representations of the Mithraic Bull on gems of four or five centuries before Christ prove that Taurus was at that time still prominent in the astronomy and religion of Persia and Babylon.

The Egyptians regarded Taurus as the emblem of a perpetual return to life. They identified it with Osiris, the Bull-god, the god of the Nile, and worshipped it under this figure by the name "Apis."

Plunket considers that the Apis Bull of Egypt was looked upon as a living representation of the zodiacal Bull, and that this figure may have been known before the building of the Great Pyramid.

The Persians also were worshippers of the Bull. They designated the successive signs of the zodiac by the letters of the alphabet, and with them A stands for Taurus, B for the Twins, etc., clearly indicating that they considered the Bull the first sign of the zodiac. Reference to the astrological books of the Jews shows that they, too, considered Taurus the leader of the zodiacal signs.

In fact in all the ancient zodiacs that have come down to us Taurus apparently began the year, and it seems to have been regarded as a Bull in all of the ancient Mediterranean countries, and also in countries far distant from Europe, and from the scenes of Hellenic mythology.

The constellation is exceedingly rich in myth and legend. According to Grecian mythology, this is the Bull that carried Europa over the seas to that country which derived from her its name. She was the daughter of Agenor, and, it is said, so beautiful that Jupiter fell in love with her. He assumed the form of a snow-white Bull and mingled with the herds of Agenor. Europa, charmed with the sight of the beautiful creature, had the temerity to sit upon his back. The god took advantage of the situation and carried Europa across the seas to Crete.

In Moschus, translated by Andrew Lang, we read of Jupiter's achievement and of his journeyings with Europa:

Swiftly he sped to the deep . . .

The strand he gained and forward he sped like a dolphin, faring with unwetted hooves over the wide waves, and the sea as he came grew smooth, and the sea monsters gambolled around before the feet of Jupiter, and the dolphin rejoiced and rising from the deeps he trembled on the swell of the sea. The Nereids arose out of the salt waters and all of them came on in orderly array, riding on the backs of sea beasts.

Tennyson in his "Palace of Art" thus alludes to Europa:

Or sweet Europa's mantle blew unclasp'd,
From off her shoulder backward borne;
From one hand droop'd a crocus; one hand grasp'd
The mild Bull's golden horn.

The kidnapping of Europa has been a source of inspiration to a host of poets and artists in all ages. On the ceiling of the Ducal Palace in Venice there is a celebrated painting by Paul Veronese depicting the Rape of Europa.

The following sonnet by Wm. W. Story is descriptive of this picture:

Zephyr is wandering here with gentle sound
The first fresh fragrance of the Spring to seek;
The milk-white steer, whose budding horns are crowned
With flowery garlands, kneeling on the ground
Receives his burden fair, and turns his sleek
Mild head around, her sandalled foot to lick;
Luxuriant, joyous, fresh, with roses bound
About her sunny head, and on her cheek
The glow of morn, Europa mounts the steer.
One handmaid clasps her girdle, and one calls
The hovering Loves to bring their garlands near.
From her full breast the loosened drapery falls,
As borne by Love o'er slope and lea she goes,
Glad with exuberant life fresh as a new-blown rose.

Again we read:

Now lows a milk-white bull on Asia's strand,
And crops with dancing head the daisied land,
With rosy wreaths Europa's hand adorns
His fringed forehead and his pearly horns.
Light on his back the sportive damsel bounds,
And, pleased, he moves along the flowery ground,
Bears with slow steps his beauteous prize aloof,
And, dips in the dancing flood his ivory hoof.

Jupiter's exploit was commemorated on earth by the naming of a continent, and in the heavens by the constellation Taurus.

There is every reason to suppose, however, that Taurus antedated the period of Greek interest in astronomy, and that the constellation was invented by the Egyptians or Chaldeans.

With the Romans, prior to the reign of Julius Cæsar, the year began in March, when Taurus is just visible in the western horizon setting after the sun. "The white Bull opens with his golden horns the year," is the way Virgil expresses it.

The idea of whiteness in connection with Taurus seems to have had a very early origin. It probably arose from the Greek legend of the mythical Bull, which is always described as snow white.

Among the ancient Chinese Taurus was known as "the White Tiger"; later it was called "the Golden Ox." Strangely enough we find that the South American Indians of the Amazon country called this star group "the Ox." Here again is further proof that at a very early date there was a transmigration, or a means of communication unknown to us, between the far east and the far west.

Aratos refers to the Bull as "Crouching." Manilius speaks of "the striving Bull," and according to Cicero, the Bull's knees are "bent." The Bull is depicted as in a crouching attitude, in accordance with the legend, that Europa might the more easily mount upon his back. It is not clear why only half the figure is shown, when there was sufficient space and stars were not lacking to depict the entire figure. In the half horse, Pegasus, we have a similar incongruity which is difficult to explain. In the case of Pegasus, as has been explained, the horse is presumed to be flying upwards through the clouds and therefore but half of the creature appears. In like manner the Bull is supposed to be swimming and half his body is submerged.

Jensen identifies Taurus with Marduk, the Spring Sun, which was worshipped as far back as 2200 B.C. He is of the opinion that the constellation was formed as early as 5000 B.C., even before the equinox lay there.

The Bull was an important object of worship with the

Photo by Naya

The Rape of Europa

Painting by Veronese. In the Ducal Palace, Venice

Druids, and their great Tauric festival was held when the sun entered this constellation, a survival of which has come down to us in the festival of May Day.

It has been claimed, says Allen, that the tors of England were the old sites of the Tauric worship of the Druids, and our hot cross buns are the present representatives of the early bull cakes, with the same stellar association tracing back through the ages to Egypt and Phœnicia. According to a Scotch myth the Candlemas Bull is seen rising in the twilight on New Year's eve.

Mrs. Benjamin has written a most interesting monograph on the sun in Taurus which the writer takes the liberty of quoting from, as it reveals much that is enlightening concerning the constellation, and the customs that have survived the ancient worship of this time-honoured star group:

"In all ages of the world the nations have hailed with delight the return of spring and the revivification of nature under the warmth and heat of the sun. This universal festival we know as May Day and it commemorates the entrance of the sun into the constellation Taurus at the vernal equinox 4000 B.C. It is still observed in all parts of Great Britain, among us, and in India and Persia.

"The old English 'Morris Dance' is a remnant of this festival time, and Maurice says, 'I have little doubt that May Day or at least the day on which the sun entered Taurus has been immemorially kept as a sacred festival from the creation of the earth and man, and was originally intended as a memorial of that auspicious period and momentous event'."

In the Hebrew, Syriac, Arabic, and Coptic the word for bull means "coming" or "who cometh," and the lucida of the constellation is a first magnitude star called "Aldebaran," which means the "leader."

The Masonic Tau Cross, , is an expressive symbol of the vernal equinox and of immortality. The emblem is found on many of the ancient monuments of Egypt, and clearly its astronomical significance can be traced to the constellation Taurus, for Brown tells us that the word "Tau" is derived from an Egyptian or Coptic root meaning a bull or cow.

The ancient hieroglyphic sign of this constellation  represents the face and horns of a bull.

The Greek letter Tau (τ) and the English (T) are derived from this symbol by the following steps:

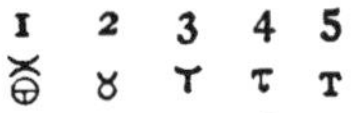

In the Hebrew zodiac Taurus is ascribed to Joseph, and Dr. Seiss asserts that Taurus represents the fabled unicorn. In the so-called "Apostolic Zodiac" Taurus was said to represent St. Andrew, or the Burnt Sacrifice.

Astrologically speaking, says Proctor, Taurus gives to its natives (those born from April 19th to May 20th) a stout athletic frame, broad bull-like forehead, dark curly hair, short neck, a dull apathetic temper, exceedingly cruel and malicious if once aroused. It governs the neck and throat, and reigns over Ireland, Poland, part of Russia, Holland, Persia, Asia Minor, the Archipelago, Mantua, Leipsic, etc. It is a feminine sign and unfortunate. The flower is the jonquil, and the stone, agate. It was considered under the guardianship of Venus, and white and lemon were the colours assigned to it.

> . . . go forth at night,
> And talk with Aldebaran where he flames
> In the cold forehead of the wintry sky.
>
> Mrs. Sigourney.

The Arabic name for α Tauri is "Aldebaran," which means the "leader," or the attendant or follower, *i. e.*, of the Pleiades. It was also known to the Arabs as "the Eye of the Bull," and "the Great Camel," "the Stallion Camel," "the Fat Camel," "the Female Camel," and "the Bull's Heart."

The Hindus called the star "Rohini," meaning the "Red

Deer," probably because of its colour, which is decidedly ruddy.

According to Lockyer, Aldebaran rose heliacally at the beginning of spring in Babylon 6900 years ago, and it was thought that its rising at this time unattended by showers portended a barren year. The Babylonians regarded Aldebaran as "the Leading Star of Stars," as it was the brightest star in the first of the zodiacal signs.

The Akkadians called it "the Furrow of Heaven," and "the Messenger of Light," although Allen tells us that this latter title was applied to Hamal, Capella, and Vega.

Astrologically Aldebaran was a fortunate star, portending riches and honour, and it was one of the four "Royal Stars" or "Guardians of the Sky" of Persia, 5000 years ago, when it marked the vernal equinox.

Mrs. Martin sees in these four starry Guardians of the Sky a suggestion of royalty: "As one slips away from our admiring gaze we turn to hail the coming of the other. 'The King is dead: long live the King.'" The rising of Aldebaran is thus described by Mrs. Martin: "Along in September a very little north of east it shows its fiery face above the horizon with such unmistakable individuality that it catches the eye of even the least observing. . . . It glows with a rosy light that demands recognition and at once pronounces it one of the most important heavenly bodies."

According to Peschitta the line in the book of Job, "Dost thou guide ʿAyish and her children?" refers to Aldebaran and the Hyades. "ʿAsh" means "moth" and the Hyades are V-shaped, resembling a butterfly or moth.

Aldebaran lies along the moon's track and is often occulted by our satellite. Because of its position it is a star much used by navigators in ascertaining their position. It is nearly a standard first magnitude star, lacking only two tenths of a magnitude of so being. Elkins states that Aldebaran is twenty-eight light years distant from us.

This enormous distance is perhaps better guaged when we say that if the distance from the earth to the sun, a matter of ninety-three million miles, be considered as one inch, Aldebaran would be twenty-seven miles away.

Aldebaran is said to be receding from us at the rate of thirty miles a second, and Prof. Russell tells us that this gigantic sun emits 160 times as much light as our sun. It culminates at 9 P.M., Jan. 10th.

β Tauri, also known as γ Aurigæ, a second magnitude star, was called by the Arabs "El-Nath," which means the "Butting One," a reference to its position in the northern horn of the Bull. This star is common to the constellations Taurus and Auriga.

Aratos thus refers to it:

> The tip of the left horn and the right foot
> Of the near Charioteer, one star embraces.

The star is peculiarly white in colour, and Allen tells us that "the sun stood near this star at the commencement of spring 6000 years ago. It has a Sirian spectrum, and is receding from us at the rate of about five miles a second. Between it and ψ Aurigæ was discovered on the 24th of January, 1891, the now celebrated Nova Aurigæ that has occasioned so much interest in the astronomical world."

Among the Hindus it represented Agni, the god of fire, and among the astrologers it portended eminence and fortune.

ζ Tauri, a 3.5 magnitude star, marks the tip of the southern horn of the Bull. The wonderful "Crab Nebula" is situated a little north-west of it, and can be seen in a three-inch glass, though a powerful telescope alone reveals its curious form.

Astrologically ζ Tauri was considered of mischievous influence.

Taurus contains the greatest number of stars of any constellation, 141 in all, exclusive of the Pleiades.

The celebrated star clusters, the Hyades and Pleiades, are contained in Taurus. As they are specially noteworthy the writer has seen fit to devote a chapter to them.

Ursa Major
The Greater Bear

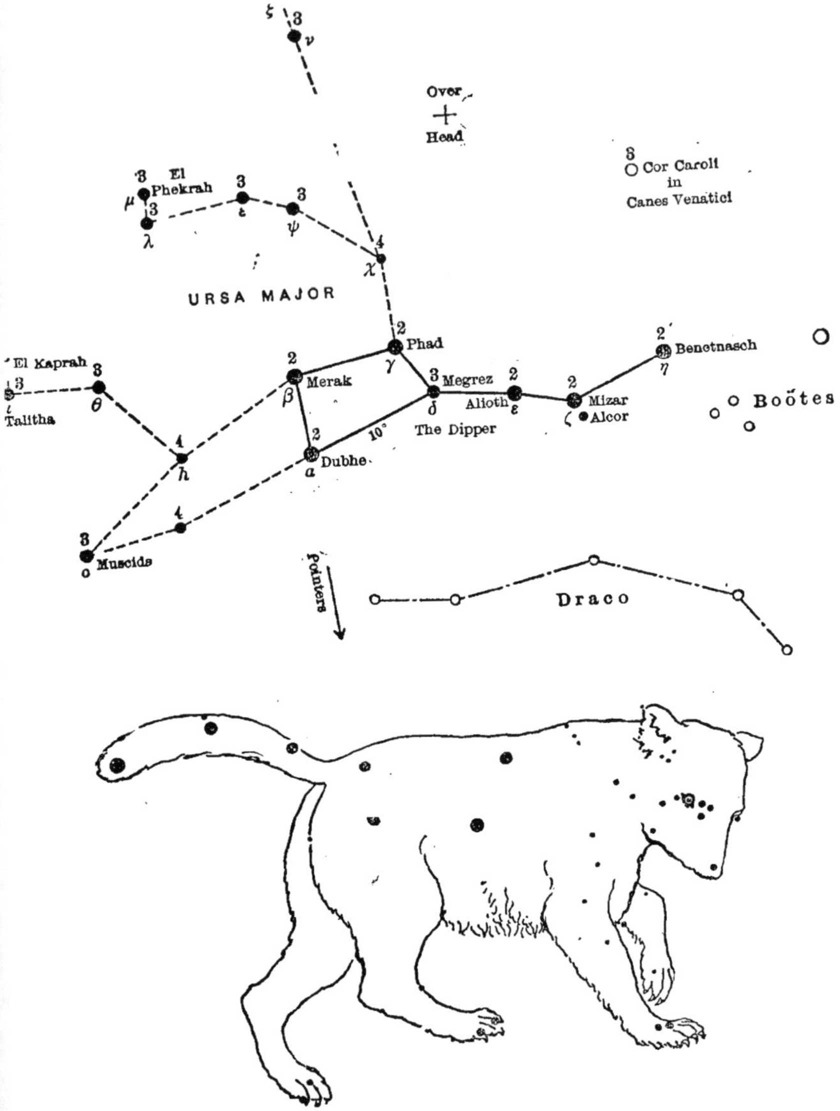
Over
Head
Cor Caroli
in
Canes Venatici
El
Phekrah
URSA MAJOR
El Kaprah
Talitha
Phad
Merak
Megrez
Alioth
Benetnasch
Mizar
Alcor
Boötes
The Dipper
Dubhe
10°
Muscids
Pointers
Draco

URSA MAJOR

URSA MAJOR
THE GREATER BEAR

He who would scan the figured skies
  Their brightest gems to tell
Must first direct his mind's eye north
  And learn the Bear's stars well.

URSA MAJOR, or the Greater Bear, is the most easily recognised and the most widely known of all the constellations. In all the records of an astronomical character that have come down to us we find allusions to this famous group of northern stars. It is unquestionably the most ancient of all the constellations, and universally known as "the Bear."

On the banks of the Euphrates thousands of years ago it was so designated, and the Iroquois Indians of North America called this star group "Okouari," their name for "bear." The Algonquin Indians called the constellation "the Bear and the Hunters," and as they were evidently sensible of the incongruity of attributing a conspicuously long tail to an animal that had none, they consequently regarded the three stars in the tail of the Bear as three hunters pursuing the beast.

The Finns called Ursa Major "Otawa," a title resembling the "Okouari" of the Iroquois, and it is inferred that they regarded this constellation as representing a Bear.

Thus in remote parts of the earth, in the far north, from the valley of the Euphrates to the region of the Great Lakes of North America, we find the same stars likened to an identical animal, "the relic of some primeval association of ideas long since extinct."

The arrangement of the stars in Ursa Major in no way

suggests a bear, or any other animal, and even if one nation should so picture it, there is no reason why the same imaginative creation should be universally identical.

Aristotle held that the name was derived from the fact that of all known animals the bear was thought to be the only one that dared to venture into the frozen regions of the north and tempt the solitude and cold.

Prof. Max Müller thinks that the name of the Great Bear is the result of a mistake as to the meaning of words. The Sanscrit name "Riksha" signifies both "bear" and "star that is bright." The seven bright stars in this constellation form such a striking group that they might well merit the title "Riksha" in its latter sense. It has been suggested that the constellation was called "the Bear" as a pun on this word "Riksha." Later on, this word was confounded with the word, "Rishi," and so connected with the Seven Sages or Poets of India, the Seven Wise Men of Greece, the Seven Sleepers of Ephesus, and the Seven Champions of Christendom. The ancient Hindus believed that the seven bright stars of Ursa Major represented the seven principal rich men or holy persons who were supposed to live beyond Saturn, but the inhabitants of northern Asia, the Phœnicians, Persians, and others, all saw in these bright stars of the north the likeness of a great bear.

On the famous zodiac of Denderah on the Nile, is pictured the leg of an animal. This is identified by the authorities with a constellation called "the Thigh," which beyond question is the figure now known to us as Ursa Major. The Egyptians called this constellation "the Hippopotamus," "the Dog of Set," or "of Typhon," and in latter days "the Car of Osiris."

The Greeks called this star group Ἄρκτος μεγάλη, whence we get our word "arctic."

It has been suggested that the word "Ursa" is derived from "Versus," because the constellation is seen to turn about the Pole.

Homer mentions the Bear as keeping watch from his arctic den upon the hunter Orion for fear of a sudden attack. He regarded the constellation as only composed of the seven stars which form the familiar figure of the Dipper, and in his description of the shield of Achilles, he writes, after mentioning other stars, of "the Bear surnamed the Chariot." Homer's twice repeated assertion that "the constellation of the Bear alone never sinks into the ocean" merely allows us to infer that in his age the Greek sphere did not yet comprise the constellations Draco, Cepheus, Cassiopeia, and Ursa Minor, which likewise never set.

Even in Homer's day Ursa Major was known as "the Wain," the name by which it is known in England today. This title was originally "Charlemagne's Wain," from the Scandinavian Karlsvagn, the Carle's Wain. Another title was Arthur's Wain, a name, says Smyth, derived from the Welsh "Arth," a bear. Smyth finds in the circling of this constellation about the Pole the possible origin of King Arthur's famous Round Table.

In all probability it is this group of stars and not Arcturus which is referred to in Job's question:

> Canst thou guide Arcturus with his sons?

In the Revised Version it reads:

> Canst thou guide the Bear with her train?

The word from which Arcturus was derived was "ʿAyish" or "ʿAsh," a word that does not differ importantly from the word "naʿsh," the Hebrew word for assembly, the Arabic "bier," a title among the Arabs for the four stars forming the Dipper, from remote antiquity.

The three stars which form the tail of the Bear were called by the Arabs "Benāt-naʿsh," the "daughters of the Bier." "Regarding Arcturus as referring to the Bear," says Maunder, "we have in both passages of Job which mention Arcturus, Orion, the Pleiades, and the Chambers of the South the four quarters of the heavens marked out as being under the dominion of the Lord. In the ninth chapter they are given in this order: The Bear which is in the north, Orion in its acronical rising with the sun setting in the west, the Pleiades in their heliacal rising with the sun rising in the east, and the Chambers of the South." In the Breeches Bible the note on the word "Arcturus" reads: "The North star with those that are about him."

It seems more consistent with the stellar arrangement to regard the four stars forming the bowl of the Dipper as representing a bear, and the three stars in the handle of the Dipper as representing the cubs following in her steps, "her train," than to regard the constellation as a bear with a long tail.

The Arabs also had a consistent figure in the Bier with three mourners following. This title "the Bier" is so similar to the almost universal appellation "the Bear," that we might almost suppose that the latter title was a confused rendering of the former.

In some time antedating history, nomads of the east familiar with this constellation of "the Bier" may have reached North America and there conveyed their conception of this star group to the Indians, who translated the idea into terms familiar to their lives. The bier was distinctly an object familiar in the Orient, and foreign to the western savage, whereas the bear was foreign to the far east and familiar to the western Indian, whose life was bound up in the hunt. It therefore seems natural that in the east we should find the bier followed by the mourners represented by these prominent stars of the northern sky, and among the Indians we should see this same star group likened to a bear pursued by hunters.

Proctor is of the opinion that originally this was the only Bear constellation, and that it was a much larger figure than at present. Modern astronomers, finding a great vacant space where formerly the Bear's large frame extended, formed there the new constellation "Canes Venatici," the Hunting Dogs. No one can recognise a bear in the present figure of the constellation, but Proctor says that one who looks attentively at the part of the skies occupied by the constellation will recognise, if they are imaginative, a monstrous bear with the proper small head of creatures of the bear family, and with exceedingly well developed plantigrade feet.

The feet of the Bear are marked by three pairs of stars strikingly arranged, and Maunder agrees with Proctor in considering that these conspicuous stars suggested the feet of a great plantigrade animal. Of course the figure cannot at all times be recognised with equal facility, but before midnight during the last four or five months in the year the Bear is seen, either upright in the heavens, or as if descending a slope, and favourably situated for observing.

Stories of the descent of tribes from animals are widespread among the ancient annals of the race, the Akkadians, Australians, red Indians, Bushmen, Bedouins, and other wild races believing that they sprang from such a source. The Akkadians considered that they were descended from a bear, and hence transferred the creature in fancy to the stars. They were known among the ancients as "the Bear Folk."

The growth of the Bear from his original seven stars was obviously prompted by a desire to make the animal correspond in size to the long tail which appeared in the original figure. The stars adapted themselves fairly well for the purpose, and there was no other constellation in the way.

The Tower of Babel, the most ancient of temples, was called "the Temple of the Seven Lights," or "the Celestial Earth." It embodied the astronomical kingdoms of antiquity. The seven lights were, it has been thought, the seven stars of the Great Dipper.

Mythology links together in one story the constellations of the Greater and Lesser Bears:

The legend relates that Callisto, a nymph, the beautiful daughter of Lycaon, King of Arcadia, incurred the jealous wrath of Juno. Jupiter, fearing that Callisto would suffer injury at Juno's hands, transformed her into a bear. Juno on perceiving this induced Diana to kill the bear in the chase, but Jupiter placed his favourite out of harm's way in the starry skies. Callisto's son Arcas afterwards became the constellation of Ursa Minor.

Addison, in his translation of Ovid's *Metamorphoses*, thus writes that Jove—

> snatched them through the air
> In whirlwinds up to heaven and fix'd them there;
> Where the new constellations nightly rise,
> And add a lustre to the northern skies.

Juno, it is said, indignant at the honour thus shown the objects of her hatred, persuaded Tethys and Oceanus to forbid the Bears to descend like the other stars into the sea.

Homer in the following lines thus alludes to the perpetual punishment meted out to Callisto and Arcas:

> Arctos, sole star that never bathes in th' ocean wave.

Bryant also writes in like vein:

> The Bear, that sees star setting after star
> In the blue brine, descends not to the deep.

The Bear now sets except in high latitudes, but in Homer's day and long before, his stars did not sink below the horizon or lave the seas.

Lowell in "Prometheus" thus refers to the Bear:

> One after one the stars have risen and set,
> Sparkling upon the hoar frost of my chain

Photo by Brogi

Juno and Jove
National Museum, Palermo

The Bear that prowled all night about the fold
Of the North Star hath shrunk into his den,
Scared by the blithesome footsteps of the dawn.

Ovid gives a slightly different version of the legend. According to him, Juno changed Callisto into a bear, and when Arcas was out hunting and unwittingly about to slay his mother in the guise of a bear, Jupiter placed the bear and the hunter among the stars.

According to another legend this constellation represented a Princess, transformed into a bear on account of her pride in rejecting all suitors. For this her skin was nailed to the sky as a warning to other proud maidens.

Aratos made the two Bears the Cretan nurses of the infant Jupiter, afterwards raised to heaven for their devotion to their charge. Lewis disregards this legend on the ground that Crete never contained any bears.

A modern Grecian legend relates that originally the sky was supposed to be made of glass which touched the earth on both sides. It was soft and thin, and some one nailed a bearskin upon it. The nails became stars, and the tail of the Bear is represented by three bright stars, which are also known as "the handle of the Great Dipper."

The Iroquois Indians had a legend concerning this constellation which was as follows:

"A party of hunters pursuing a bear were attacked by three monster stone giants, who destroyed all but three of them. These, together with the bear, were carried up to the sky by invisible hands. The bear is still being pursued by the three hunters. The first carries a bow, the second a kettle to cook him in (this is represented by the little star Alcor), and the third carries sticks with which to light a fire when the bear is slain. In the autumn the first hunter hits the bear, and the bloodstains from the wounded bear tinge the autumn foliage." This legend is similar to that of the Housatonic Indians, who roamed through the valley from Pittsfield to Great Barrington. They believed that the chase of the bear lasted from spring until autumn, when the animal was wounded, and its blood was seen on the crimson foliage of the forest.

Stansbury Hagar, in an interesting monograph on *The Celestial Bear*, relates much of interest in this connection, which the writer takes the liberty of quoting in part. In some particulars the legends he recites coincide with the Indian legends previously referred to, but there are many interesting details in addition which reveal the active imagination of the American Indians in its relation to these famous stars.

"It is probable that in no part of the world has the observation of the stars exerted a greater influence over religion and mythology than amongst the native civilised people of Central and South America. Throughout their mythology the most beautiful legends are those associated with the heavens, and the two stellar groups which seem to have played decidedly the most conspicuous parts in these legends are the Pleiades and the Great Bear.

"These star groups figured prominently in the legends of the North American Indians also, and we can easily imagine the astonishment of the early missionaries when they pointed out the stars of the Great Bear to the Algonquins and received the reply: 'But they are our Bear stars too.'

"This constellation, famous in the mythology of the Orient, seems to have been called 'the Bear' over nearly the whole of our continent, when the first Europeans of whom we have knowledge arrived.

"It was known as far north as Point Barrow, as far east as Nova Scotia, as far west as the Pacific coast, and as far south as the Pueblos. The best-known legend concerning this star group is common to the tribes of the Algonquin and Iroquois Indians, and beside Ursa Major it embraces the neighbouring constellations Boötes and Corona Borealis. It is in the form of a drama with the following Dramatis Personæ:

	The Bear........	α, β, γ, δ,	Ursæ Majoris
The Hunters	" Robin........	ε	" "
	" Chickadee....	ξ	" "
	" Moose Bird...	η	"
	" Pigeon........	γ	Boötis
	" Bluejay......	ε	"
	" Owl..........	α	" ,Arcturus
	" Saw-whet.....	η	
	" Pot..........		Alcor
	" Den.........	μ, δ,	Boötis and the bright stars in Corona Borealis

The Bear is thus represented by the four stars in the bowl of the Dipper, and behind are seven hunters pursuing her. Here again we find the number seven associated with this constellation.

"The first hunter was called 'the Robin' because that star has a reddish tinge, the second 'the Chickadee' because its star is smaller than the others, the fifth hunter the Bluejay because its star is blue. Arcturus becomes the Owl because of its large size, and the star of the seventh hunter, the Saw-whet, because its reddish hue suggests the brilliant feathers which mark the head of that bird.

"Close beside the second hunter is a little star (Alcor), which represents the Pot which he is carrying to cook the bear meat in. Just above the hunters is a group of smaller stars which represent the Bear's den.

"Late in the spring the Bear, waking from her long winter sleep, leaves her rocky hillside den and descends to the ground in search of food. Instantly the sharp-eyed Chickadee perceives her, and being too small to undertake the pursuit alone, calls the other hunters to his aid. Together the seven start after the Bear, the Chickadee with the Pot being placed between two of the larger birds so that he may not lose his way. All the hunters are hungry and pursue eagerly, but throughout the summer the Bear flees across the northern horizon and the pursuit continues. In the autumn one by one the hunters in the rear begin to lose their trail. First of all the Owl, heavier and clumsier of wing than the other birds, disappears from the chase, next the Bluejay and Pigeon also lose the trail and drop out. This leaves only the Robin, Chickadee, and Moose Bird. At last about mid-autumn they succeed in overtaking their prey. The Bear at bay rears up and prepares to defend herself, but the Robin pierces her with an arrow and she falls over on her back. The Robin, in haste to feed upon the Bear, leaps upon his victim and becomes covered with blood. Flying to a maple tree near at hand in the land of the sky, he tries to shake off the blood, and succeeds in getting it all off save a spot upon his breast. 'That spot,' says the garrulous Chickadee, 'you will carry as long as your name is Robin.' The blood which the Robin shook off spattered far and wide over the forests of earth below, and hence we see each autumn the blood-red tints on the foliage. The Chickadee now arrives on the scene and with the Robin cuts up the Bear, builds a fire, and cooks the meat. The Moose Bird now appears; he knew the others would catch the Bear and prepare the meat, and wanted only to be on time to share it, so whenever a bear or a moose or other animal is killed to-day you will see him appear to demand his share. That is why he is called 'He-who-comes-in-at-the-last-moment.'

"Through the winter the Bear's skeleton lies upon its back in the sky, but her life spirit has entered into another Bear who also lies upon her back in the den, invisible, sleeping the winter sleep. When the spring comes around, the Bear will again issue forth from the den, to be again pursued by the hunters, and so the drama keeps on eternally."

With the Zuñis, Ursa Major was important as marking the seasons. They say that when winter comes the Bear lazily sleeps, no longer guarding the westland from the cold of the ice gods and the white down of their mighty breathing, but, when the Bear, awakening, growls in the springtime and the answering thunder mutters, the strength of the ice gods being shaken, the reign of summer begins again.

The Chinese say that in spring the tail of the Bear (the Micmac three hunters) points east, in summer south, in autumn west, in winter north,—a correct statement for the forepart of the evening.

The Ojibway Indians have a legend which relates that a southern star came to earth in the form of a beautiful maiden, bringing the water lilies. Her brethren can be seen far off in the north hunting the bear, whilst her sisters watch her in the east and west.

The Onondaga Indians knew the stars representing the bear's den, which is formed by the stars in the constellation Corona Borealis. The Cherokees also know the legend of the celestial bear hunt, and say that after the three hunters have killed the bear in the fall they lose the trail and circle helplessly around till spring. They assert that the honey dew, which is noticeable in the autumn, comes from the bear's fat which they are trying out over a fire.

The Blackfeet Indians have known these seven stars of the Dipper as seven boys, all of whom had been killed by their sister save the youngest (the star Dubhe), who killed her in turn.

The Point Barrow Eskimos recognised the stars of the Bear with the seven hunters around him, and the Zuñis call the group "the Great White Bear of the seven stars." These stars seem to have played an important part in Pueblo mythology.

The Thlinkeets of the Pacific Coast regarded the stars of Ursa Major as representing a Bear. They thought that the Bear was so-called because its stars act so like a bear, slowly circling about; then too there is a den represented by a group of stars, and no other animal save the bear has a den of that shape. The Micmac Indians noticed these similarities between the position of these stars and the habits of the bear, and they were the source of many of the Indian legends.

According to a Basque legend, a farmer had two of his oxen stolen by two thieves. He sent his servant in pursuit of them, and as he did not return he despatched his housekeeper and dog, and finally as no one returned he went after the thieves himself. Because he lost his temper in his search for the oxen he is condemned to continue it for ever, and thus we find them all represented in the seven stars of the Dipper. The first two stars (the Pointers) represent the two oxen, then follow the two thieves, the servant, the housekeeper with the little dog (the star Alcor), and lastly the farmer himself.

The Basques are also said to believe that when the Bear is above the Pole the season is hot and dry, when below it the season is wet.

Another legend respecting these famous stars relates that they represent a peasant's waggon. The peasant, so the story runs, met our Saviour near the shores of Galilee, and gave him a ride in his waggon. He was rewarded for his kindness by a place in the heavens, whither he and his conveyance were transported.

To the Eskimos, Ursa Major represented four men carrying a sick or dead man. The idea of a bier associated with the constellation in the east seems to be embodied in this notion of the Eskimos. The Eskimos also recognised the Great Dipper as a herd of reindeer.

Ursa Major was used long before the invention of the mariner's compass to guide the paths of ships at night, as Manilius informs us:

> Seven equal stars adorn the greater Bear
> And teach the Grecian sailors how to steer.

These stars were equally valuable as guides to those who travelled long distances through unknown lands.

According to Diodorus Siculus, travellers in the sandy plains of Arabia were accustomed to direct their course by the Bears.

The Greeks made the Great Bear their guide in navigation, whereas the Phœnicians steered by the Lesser Bear.

The Greeks called the Great Bear καλλίστη from the Phœnician "kalitsah," meaning safety, as the observation of these stars helped to a safe voyage.

Aratos wrote:

> By it on the deep
> Achaians gather where to sail their ships.

In the *Odyssey*, the sailing directions to Ulysses bid him keep the Bear always on the left, that is, to steer due east.

Aratos says that the Sidonians steer by the Little Bear, and that it is preferable to the Great Bear as it is situated nearer the Pole.

In this connection Apollonius mentions Ursa Major, which was often called "Helice" by the Greeks:

> Night on the earth pour'd darkness on the sea,
> The watchful sailor to Orion's star
> And Helice turned heedful.

Among the Chinese, the Great Bear was known as a bushel or measure of corn, the tail being the handle of the measure. Here, as in the case of the titles "Wain," "Plough," and "Bier," we have a plain case of imitative name giving.

The ancients associated the idea of dancing with Ursa Major and the other circumpolar constellations, and they not infrequently mention "the dances of the stars." The two Bears were imagined as reeling around the Pole like a pair of waltzers.

> Onward the kindred bears with footsteps rude
> Dance 'round the pole, pursuing and pursued.

This comparison is drawn from the circular dances of

the Greeks, and alludes principally to the motion of the stars in the immediate vicinity of the Pole.

> . . . round and round the frozen pole
> Glideth the lean white Bear.
>
> Buchanan.

There is some little interest in Ursa Major on account of the possibility of its being used as a kind of celestial timekeeper. The northern sky is in reality a great clock dial, over which hands wrought of stars trace their way unceasingly. Moreover, it is a timepiece that is absolutely accurate, and which requires no winding or repairing. A line drawn through α and β Ursæ Majoris, or "the Pointers" as these stars are called, passes almost exactly through the pole of the heavens. Now this line revolves with the constellation once in twenty-four hours. On March 21st at 10.55 P.M., the superior passage takes place; a like passage, but invisible, occurs on Sept. 22d at 10.55 A.M. Knowing the day of the month, the time may be derived by observing what angle the line joining these stars makes with the vertical. In Shakespeare's *King Henry IV.* the Carrier exclaims:

> Heigho: an't be not four by the day I 'll be hanged
> Charles's Wain is over the new chimney.

And Falstaff says:

> We that take purses go by the moon and the seven stars.

Poe in one of his poems writes:

> And star dials pointed to morn.

Tennyson wrote:

> We danced about the May-pole and in the hazel copse
> Till Charles's Wain came out above the tall white chimney tops.

And again in *The Princess:*

> I paced the terrace till the Bear had wheel'd
> Thro' a great arc his seven slow suns.

Spenser also alludes to this celestial timepiece in the *Faerie Queene:*

> By this the northern wagoner had set
> His sevenfold time behind the steadfast starre.

In a Blackfoot Indian myth we read: "The seven Persons [the Dipper] slowly swung around and pointed downward. It was the middle of the night." This shows that the Indians were accustomed to mark time at night by the position of the circumpolar stars.

Allen tells us that the Bears have been frequently found on the old signboards of English inns, and in a more important way are emblazoned on the shields of the cities of Antwerp and Gröningen, in the Netherlands.

The well-known talisman of good luck, the Swastika Cross, is considered the oldest cross and symbol in the world. It is said to have been familiar to primitive man as a part of the constellation of Ursa Major, the portion popularly known as "the Dipper." The stars that trace the cross form the figure of a Dipper whichever way the cross is turned.

The Arabs imagined Ursa Major and Ursa Minor to be a gazelle and its young, and the three conspicuous pairs of stars in the feet of the Great Bear represented to them the footprints of several gazelles, which, according to a legend, sprang from that spot when the Lion lashed the sky with his tail. The Lion, so the story runs, pursued the gazelles, and some of them jumped for safety into the Great Pond which is formed by a group of stars in Ursa Major.

α Ursæ Majoris is named "Dubhe," meaning the "Bear" or "She Bear." The title is derived from an Arab phrase meaning the back of the Bear. Lockyer identifies this star with the Egyptian "Āk," meaning the "Eye," the prominent one of the constellation. This star was utilised in the alignment of the walls of the temple of Hathor at Denderah, and was the orientation point of that structure before 5000 B.C.

The Chinese called this star "Heaven's Pivot." It is located five degrees from β and ten degrees from δ Ursæ Majoris, and about twenty-eight degrees from Polaris, the Pole Star. These measures are useful to bear in mind in estimating celestial distances.

α and β Ursæ Majoris have been called "the Pointers," "the Keepers," and "the Two Stars." Dubhe is the only star in the Dipper that is of the solar type. It is approaching our system at the rate of twelve miles a second, and has an 11th magnitude companion discovered by Burnham in 1889.

β Ursæ Majoris was known to the Arabs as "Merak," meaning the "loin." The Chinese called it "an armillary sphere," and the Hindus regarded it as "Pulaha," one of the Rishis. It is of the Sirian type, a spectroscopic double, and is approaching the earth at the rate of eighteen miles a second.

γ Ursæ Majoris, also called "Phecda" or "Phad," meaning the "Thigh," is approaching our system at the rate of sixteen miles a second.

δ is known as "Megrez," meaning the "Root of the Tail." It is the faintest of the seven stars in the Dipper. The position of Megrez and the star Caph, β Cassiopeiæ, is peculiar. These stars are both in the equinoctial colure, one of the great circles passing through the poles, and are almost exactly opposite each other, and equally distant from the Pole. Megrez is on the meridian at 9 P.M., May 10th.

These four stars forming the bowl of the Dipper were called by the Arabs "the coach of the children of the litter." They form the hind quarters of the Bear, the frame of the Bier, the Plough, and the Wain.

ε Ursæ Majoris bears the name "Alioth." According to Gore, this is a corruption of an Arabic word meaning "the Gulf." Alioth is approaching us at the rate of nineteen miles a second, and very nearly marks the place of the radiant point of the Ursid meteor shower of Nov. 30th. It is a spectroscopic binary.

Photo by Morgan P. Taylor

The Temple of Hathor at Denderah

ζ, also called "Mizar," is the most interesting of all the stars of the Dipper. Maunder says that in every way it is the first of double stars. The fourth magnitude star Alcor forms with it a naked eye double, and it has a closer companion visible in the telescope. Mizar was the first double star discovered telescopically, Riccioli having made the discovery at Bologna in 1650. It was also the first double star to be photographed, and the first star discovered to be double by the spectroscope. In India, Mizar was regarded as one of the seven sages. It is approaching our system at the rate of nineteen miles a second.

η Ursæ Majoris, the last of these seven famous stars, was called "Benatnasch," meaning the "Governor of the Daughters of the Bier," *i. e.*, the chief of the mourners. It was also known as "Alcaid." In China, this star was called "a Revolving Light," and it marks the radiant point of the Ursid meteors of Nov. 10th. It is approaching the earth at the rate of sixteen miles a second.

Alcor is the name of the naked eye star close to Mizar. These two stars were called "the Horse and the Rider." In North Germany the Rider is supposed to start on his journey before midnight, and to return twenty-four hours later, his waggon turning around with a great noise. The Arabs called this star "Suha," meaning the "Forgotten," "Lost," or "Neglected One," and they also called it "the Test," an allusion to its visibility, as those who could see it were supposed to be keen of sight. The Arabs had the following proverb concerning this star:

> I show him Suha and he shows me the moon.

The Arabs also called this star "Winter," and "the Little Letter."

The Greeks identified Alcor with the lost Pleiad Electra, who had wandered away from her companions and had been changed into a fox. A Latin title for the star was "the Little Starry Horseman." In England it is called "Jack on the Middle Horse," and in Germany it represented Hans the Waggoner rewarded for assisting our Saviour.

In the field with Mizar and Alcor is the so-called "Sidus Ludovicanium," an eighth magnitude star of a bluish colour. This was first observed by Einmart of Nuremberg in 1691, and in 1723 another German, thinking he had discovered a planet, named it after his sovereign Ludwig V.

There are two noted stars in this constellation that remain to be mentioned. They are known as 1830 Groombridge, and Lalande 21,185. The former has been called "the Flying Star," or "Runaway Star," from the fact that its proper motion is swifter with one exception than any other star in the heavens. Its speed is so great that it would show a displacement equal to about one third of the apparent diameter of the moon in one hundred years. According to Argelander, its pace will carry it around the entire sphere in 185,000 years. Another authority asserts that in 6000 years it will reach the asterism known as Coma Berenices. Its estimated speed is two hundred miles a second, a pace that Newcomb claims is uncontrollable by the combined attractive power of the entire sidereal universe.

According to Prof. Young this star is 37.5 light years distant.

Lalande 21,185 is noted as being the nearest star to the earth of all the northern stars. Its magnitude is 7.4, and it is estimated to be 7.5 light years distant from the earth.

Five of the seven stars in the Dipper, those from β to ζ inclusive, are moving together in the sky all very nearly parallel to the line joining the first to the last. The remaining stars, α and η, are moving in almost an opposite direction, and both are receding at almost the same rate from a point in the sky not far from Vega. The accompanying diagram illustrates this movement. See p. 367.

Because of this drift, says Flammarion, they will form

the figure of an exaggerated steamer chair 50,000 years hence, as they did a magnificent cross 50,000 years ago.

Since these stars are apparently getting farther apart, they must be approaching us, a fact which the spectroscope reveals. Their rate of speed varies from seven to ten miles a second.

These stars are all about the same distance from us, between ninety and one hundred light years, although one authority places the distance as high as 192 light years. According to Ludendorff all of the seven stars exceed our sun in brilliancy from thirty to one hundred and twenty times.

θ Ursæ Majoris is a double star, with six other stars near by in the throat, breast, and fore legs of the Bear. It describes a semicircle of stars which the Arabs called "the Throne of the Mourners." This space was also known as "the Pond," already referred to, into which the gazelles sprang when pursued by the Lion.

ι, was called "Talitha," and in China ι and θ were known as "the High Dignitary." Holden says that the companion of ι is supposed to be a planet.

λ, and μ were known respectively as Tania Borealis and Tania Australis. They mark the Bear's left hind foot, and were the Arabs' "Second Spring," *i.e.*, of the gazelle. In China they were "the Middle Dignitary."

ν and ξ mark the right hind foot of the Bear. They were the Chinese "Lower Dignitary." The latter star was the first binary of which the orbit was computed, says Allen. Savary in 1828 announced its period as sixty-one years, and this star has already made more than a complete revolution since its discovery.

ο, the star that marks the nose of the Bear, was called "Muscida," a word Allen claims was coined in the Middle Ages for the muzzle of an animal.

A few degrees from β is situated the so-called "Owl Nebula." In Lord Rosse's sketch of it there is a striking resemblance to a skull, there being two symmetrically placed holes in it, each of which contained a star before 1850. Since that date only one star is visible.

Twenty stars in this constellation have received individual names,—evidence enough, says Allen, of its antiquity and popularity. The following is a partial list of the titles conferred on this celebrated star group by the various peoples of the earth from remote antiquity.

Ursa Major was known:

In the Euphratean Star List—as the Lord of Heaven.

On the Assyrian Tablets—as the Long Chariot.

In Egypt—it was the Thigh, Bull's Thigh, Fore Shank, Dog of Set, the Hippopotamus, and in later days the Car of Osiris.

The old Hindus called it the Seven Rishis or Wise Men.

In India—it was the Seven Bears, Seven Antelopes, Seven Bulls, Great Spotted Bull, and the abode of Seven Poets or Sages who entered the ark with Minos.

In China—it was the Ladle, the Bushel, the Government, the Divinity of the North, the Corn Measurer.

In Greece and Babylonia—the Chariot, the Plough, Helice, and Callisto.

The Christian Arabs knew the four stars in the bowl of the Dipper as the Bier or Great Coffin,—the three stars in the handle were the daughters of the Bier, Mary, Martha, and their maid.

In Rome—it was the Triones, Septentriones.

To the Hebrews—it was a Bier.

To the Syrians—it was a Wild Boar.

To the Druids—it was Arthur's Chariot.

The Seven Stars have also been known as the Seven Wise Men of Greece, the Seven Sleepers of Ephesus, and the Seven Champions of Christendom; the Butcher's Cleaver, the Big Dipper, the Brood Hen, and the Screw.

In the Middle Ages, Ursa Major was regarded by some as one of the bears sent by Elisha the prophet to devour the mocking boys, others thought it represented the Chariot of Elias.

Dr. Seiss regards it as symbolical of the heavenly sheep-fold, and Schiller figured the Bear as the archangel Michael, and Peter's Skiff.

In America—it was the Seven Little Indians, the Dipper.

In England—Charles's Wain, the Plough.

In early England—Arthur's Wain.

In Ireland—King Arthur's Chariot.

In France—the Saucepan, Great Chariot, David's Chariot.

In Italy—the Car of Boötes.

In Denmark, Sweden, Norway, Iceland, Scandinavia—Thor's Waggon, Waggon of Odin.

In Lapland—the Reindeer.

The Eskimos' name for the three stars in the tail of the Bear was the Many Reindeer.

THE SWASTIKA CROSS.

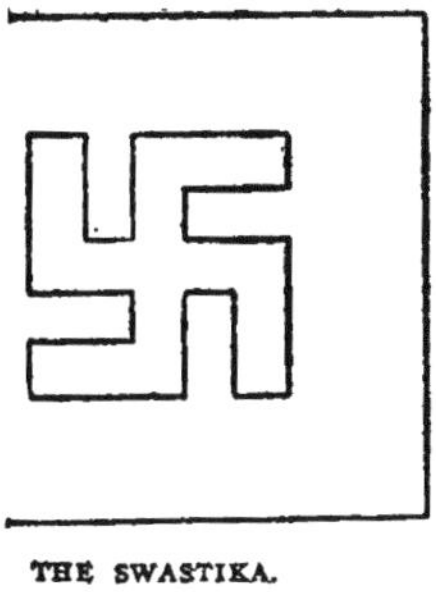

THE SWASTIKA.

PART OF THE CONSTELLATION URSA MAJOR.

Ursa Minor
The Lesser Bear

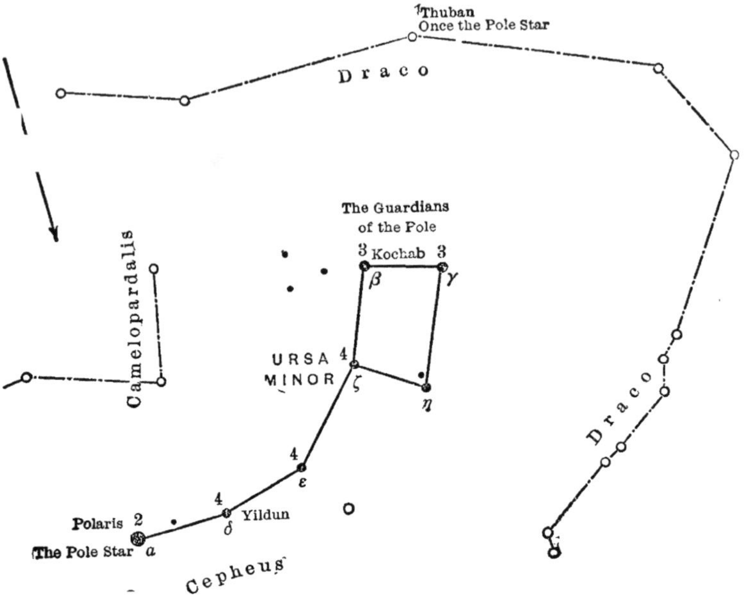
Thuban
Once the Pole Star
Draco
The Guardians
of the Pole
Kochab
Camelopardalis
URSA
MINOR
Draco
Yildun
Polaris
The Pole Star
Cepheus

URSA MINOR

URSA MINOR
THE LESSER BEAR

The lesser Wain
Is circling round the polar star.

TENNYSON.

URSA MINOR, in its present form, seems to have originated with the Phœnicians. It was not mentioned by Homer or Hesiod, for, according to Strabo, it was not admitted among the constellations of the Greeks until about 600 B.C., when Thales, inspired by its use in Phœnicia, suggested it to the Greek mariners in place of the Great Bear which hitherto had been their guide in navigating the seas. Hence the designation of the group as "Phoinike."

Observing this Phœnicians plough the main.

Aratos.

Thales is reported to have formed it by utilising the ancient wings of Draco.

The Greeks knew the constellation as "Cynosura" or "the Dog's Tail"; possibly it resembled in part the upturned coil of the tail of a dog, although one authority claims it is in no way associated with the Greek word for dog.

Brown asserts that the word is not Hellenic in origin, but Euphratean. He mentions an early constellation as "Annasurra," meaning "high in rising," certainly an appropriate title for this constellation.

Plutarch claims that the names of the Bears are derived from the use that they were put to in navigation. He says that the Phœnicians called the constellation that guided them in navigation "Doube," that is, "the speaking constellation," and that this same word happens to mean in that language a "bear," and so the name was confounded.

Aratos expressly states that the Greeks still (270 B.C.) continued to steer by "Helice" (Ursa Major), while the expert Phœnicians directed their course by "Cynosura" (Ursa Minor).

Jensen identifies this constellation with "the Leopard" of Babylonia, while on the Nile it was known as "the Dog of Set." The figure of a jackal, which is identified with this constellation, appears on the round zodiac of Denderah. The jackal also appears in the carvings on the walls of the Ramesseum.

Cæsius thought that Ursa Minor represented the chariot sent by Joseph to bring his father down into Egypt, or that in which Elijah was carried to heaven, or the bear which David slew.

According to mythology, the Bears were transferred to heaven as a reward for hiding Zeus in Crete from his cannibal father Kronos.

Ursa Minor was also identified with Arcas the son of Callisto, transported to the skies as he was about to slay his mother in the guise of a bear.

The two Bears were also fabled to have nursed Zeus on Mt. Ida. Zeus, as a reward for their faithful service, changed them into nymphs and placed them among the stars.

> The Little Bear that rocked the mighty Jove.
>
> Manilius.

The American Indians had a legend respecting this constellation which is as follows: "A hunting party of Indians lost their way, and being in doubt which way to proceed they prayed to the gods to direct them homeward. During their deliberations a little child appeared in their midst and proclaimed herself to be the spirit of the Pole Star and their guide. Following her they reached home safely, and thereafter called the Pole Star 'the star which never moves.' When the hunters died they were carried up into the heavens, and we can see them in the stars of the Little Dipper following the Pole Star faithfully every clear night."

One of the Western Indian tribes regarded Ursa Minor as a Bear, the head of the beast being represented by the three stars forming a triangle, and its back by seven other stars.

The Eskimos thought that this constellation represented four men carrying a sick baby.

Ursa Minor's chief claim to recognition lies in the universal observation of its lucida, the standard second magnitude star Alpha, known as "the Pole Star" or "Polaris," and to the Greeks as "Phœnice."

This famous star, which has been called "the lovely northern light," is the "most practically useful star in the heavens." It is the best known and most celebrated of all the stars.

The mariners of the ancient and modern worlds have placed an equal faith in the guiding beams of this steadfast star. Phœnician barks and Roman triremes, the ships of the Spanish Armada, and those that bore the early adventurers and explorers over the unknown seas, as well as the canoes and rude dugouts of the savages of many lands, have all turned their prows alike in answer to its beckoning light.

Dryden thus describes the infancy of navigation:

> Rude as their ships were navigated then,
> No useful compass, no meridian known,
> Coasting they kept the land within their ken
> And knew no north but when the Pole Star shone.

The antiquity of the knowledge of this star is attested by the fact that on the Assyrian tablets we find the Pole Star mentioned. The fact that it appears fixed was perhaps the first discovery made in the stellar universe.

It was called "the Judge of Heaven," and "the High One of the Enclosure of Light." This title was also applied to Alpha Draconis, which was in very early times the Pole Star.

In China, Polaris was called "the Great Imperial Ruler of Heaven," and "the Emperor of Emperors," and it has been from ancient times an object of worship in that land.

The ancient Mayans of Yucatan knew it as "the North Star," "the Star of the Shield," and "the Guide of the Merchants." In the Alphonsine Tables it bears the name "Alruccabah," of uncertain origin. The Greeks called it "Cynosure," and the Romans "Cynosura." Our word "cynosure" gets its meaning from Polaris, which has always been the most observed of all stars.

The Arabic name for Polaris was "the Kid," and their astronomers called it "the star of the north." It was also known in Arabia as "the hole in which the axle of the earth was borne." There was a belief among the common people of Arabia that a fixed contemplation of this star would cure itching of the eyelids.

Poets of all nations in all periods of the world's history have sung the praises of the North Star.

Marvell writes of it:

> By night the northern star their way directs.

Thomas Moore thus refers to it:

> that star, on starry nights
> The seaman singles from the sky,
> To steer his bark forever by.

Shakespeare, Milton, Wordsworth, Rossetti, and Bryant have all alluded to Polaris in their poems.

During the Civil War, escaping slaves and Northern prisoners directed their way to a harbour of refuge and home by the friendly beams of Polaris.

The Turks call the North Star "Yilduz," the star par excellence, and have a story that its light was concealed for a time after the capture of Constantinople. In Damascus it was called "Mismār," a "needle" or "nail."

Other titles for the Pole Star are "the Chariot Star," "the Steering Star," "the Lodestar," "the Northern Axle" or "Spindle."

It is generally supposed that the North Star marks the true Pole of the earth, but in reality it is 1° 14′ distant from the true Pole. Its mean right ascension, as given by the Harvard Observatory List of Bright Stars, is 1 h 22. 6 m., consequently when the right ascension of the meridian of any place is the same, Polaris will be exactly on the meridian at that time and place, but above or below the true Pole. The approximate location of the true Pole may be found by drawing an imaginary line from Polaris to ζ Ursæ Majoris. The Pole is on this line in the direction of ζ at a distance from Polaris equal to about one fourth of the distance between the Pointer stars of the Dipper.

Polaris revolves around the true Pole once in twenty-four hours in a little circle 2.5° in diameter. Within this circle two hundred stars have been photographed. "Polaris will continue its gradual approach to the Pole till about the year 2095, when it will be only 26′ 30″ away from it. It will then recede," according to Allen, "in favour successively of γ, π, ζ, ν, and α, Cephei, and α, and δ, Cygni, and α Lyræ, Vega, when, marked by this last brilliant star 11,500 years hence, the Pole will be about fifty degrees distant from its present position and within five degrees of Vega, which for 3000 years will serve as the Pole Star of the then existing races of mankind. The Polar point will then circle past ι and τ Herculis, θ, ι, and α, Draconis, β, Ursæ Majoris, and κ, Draconis back to our Polaris again, the entire period being from 25,695 to 25,868 years according to different calculations." See accompanying diagram.

Polaris is from thirty-six to sixty-three light years distant from the earth, and is receding from our system at the rate of sixteen miles a second. Its spectrum is Sirian, and as a standard second magnitude star it furnishes a means of comparison of stellar magnitudes. It has a 9.5 magnitude companion, sometimes regarded as a test star for small telescopes. This faint star has two almost dark companions revolving around it, a fact discovered by means of the spectroscope.

Polaris is presumably about the size of the sun, and at the distance of the nearest fixed star our sun would shine as a star no brighter than Polaris. It is of interest to note in passing that the North Star is elevated as many degrees above the horizon as the observer is north of the Equator, so that if a person were to stand at the North Pole, Polaris would be directly overhead.

β Ursæ Minoris was known to the Arabs as "Kochab." They also called it "the Bright One," and "the Lights of the Two Calves." The Chinese knew it as "the Emperor." Its spectrum is solar and it is receding from us at the rate of about eight miles a second.

β and γ, were known as "the Guardians or Wardens of the Pole."

Shakespeare in *Othello* thus refers to them:

> The wind-slak'd surge, with high and monstrous mane,
> Seems to cast water on the burning Bear,
> And quench the guards of th' ever fixed pole.

These stars were also called "the Dancers," and "Vigiles."

Allen tells us that these guardian stars were used as a timepiece by the common people, in the same way that Charles's Wain was used for a like purpose, as has been referred to.

γ Ursæ Minoris is a wide double, and these stars were known to the Arabs as one star, called "the dim one of the two calves."

The stars in the vicinity of the North Pole represented to the Arabs a shepherd, who, with his dog, is supposed to be pasturing a herd of sheep. To this group belong two calves, three goats, four camels, and a foal. These animals are all in the neighbourhood of Cepheus. A single camel (represented by a star in Draco) has strayed away to pasture alone. Two jackals and several hyenas are prowling about with wicked intentions.

The four stars in the bowl of the Little Dipper serve as an excellent means of comparing stellar magnitudes. The stars are β, γ, ζ, and η, and are respectively of the second, third, fourth, and fifth magnitudes.

Names by which Ursa Minor has been known:

In the Euphratean Star List—	Circler of the Midst.
In Babylonia—	The Leopard.
In Egypt—	The Jackal of Set or Sati.
In Greece—	Cynosura.
In Arabia—	Hole bearing the earth's axle.
In India—	Mount Meru. The Seat of the Gods. Dhruva.
In Scandinavia, Denmark, Iceland—	Throne of Thor. The Smaller Chariot. The Little Waggon. The Milkmaids of the Sky.
Indians of North America—	A Bear.

The Gaels called it—Fire Tail.

It has also been called "the Little Wain" or "Chariot," "the Little Dipper," "the Little Bear," "St. Michael," and "the Waggon of Joseph."

Dr. Seiss regards it as a sheepfold, and the Arabs called the three stars in the tail of the Little Bear, "the Daughters of the Lesser Bier."

In conclusion the writer quotes in part from Bryant's beautiful "Hymn to the North Star":

> The sad and solemn night
> Hath yet her multitude of cheerful fires;

The glorious host of light
Walk the dark hemisphere till she retires;
All through her silent watches, gliding slow,
Her constellations come, and climb the heavens, and go.

. . . .

And thou dost see them rise,
Star of the Pole: and thou dost see them set.
Alone, in thy cold skies,
Thou keep'st thy old unmoving station yet,
Nor join'st the dances of that glittering train,
Nor dipp'st thy virgin orb in the blue western main.

. . . .

On thy unaltering blaze
The half-wrecked mariner, his compass lost,
Fixes his steady gaze,
And steers, undoubting, to the friendly coast;
And they who stray in perilous wastes, by night,
Are glad when thou dost shine to guide their footsteps right

Virgo
The Virgin

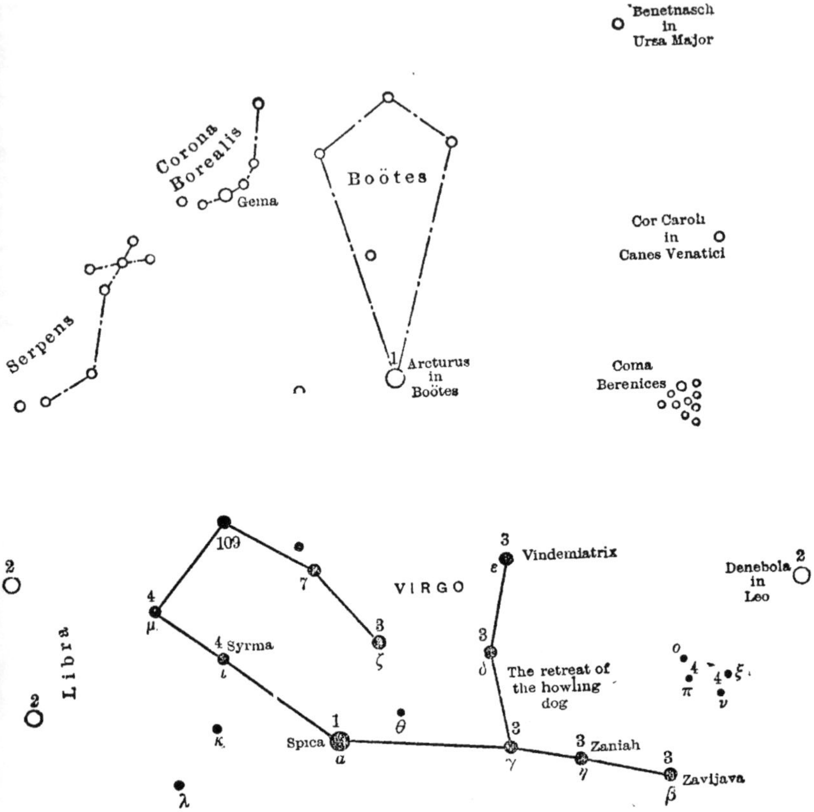
Benetnasch in Ursa Major
Corona Borealis
Gema
Boötes
Cor Caroli in Canes Venatici
Serpens
1 Arcturus in Boötes
Coma Berenices
109
7
3 Vindemiatrix
VIRGO
Denebola in Leo
Libra
4 Syrma
3 ζ
3 δ
The retreat of the howling dog
1 Spica α
θ
κ
λ
3 γ
3 Zaniah η
3 Zaviljava β
4 μ
ε
ι
ο
π
ξ
ν

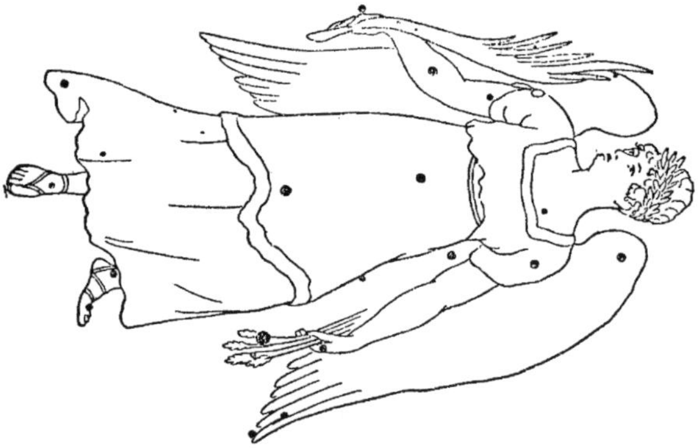

VIRGO

VIRGO
THE VIRGIN

Below Boötes thou seest the Virgin,
An ear of corn held sparkling in her hand.
Whether the daughter of Astræus, who
First grouped the stars, they say, in days of old,
Or whencesoever,—peaceful may she roll.

ARATOS.

IN the astronomical records of every age and race extant we find references to the constellation of the Virgin, and there is every reason to believe that it was one of the first star groups to receive a name.

On the ancient maps, the Virgin is generally represented as a woman with wings, in a walking attitude. In her left hand she bears a head of wheat, or ear of corn, which is marked by the brilliant first magnitude star Spica.

Her lovely tresses glow with starry light,
Stars ornament the bracelet on her hand;
Her vest in ample fold glitters with stars;
Beneath her snowy feet they shine, her eyes
Lighten all glorious, with heavenly rays,
But first the star which crowns the golden sheaf.

Brown gives us the following description of the constellation: "Virgo is the sign the sun enters in August and was depicted in the zodiac holding in her hands the emblems of the harvest. The identity of Ceres, the goddess of the harvest, with this star group is quite evident. This figure of the fruitful Virgin was placed in the zodiac as emblematic of the harvest season because the sun is in those stars at that time. The word 'Virgo' originally implied not only a Virgin but any virtuous matron. By an astronomical allegory, the Virgin of August became a goddess who descended to the earth, presided over the harvest, taught mankind agriculture, and was worshipped under various names."

Maunder does not agree with Brown's statement that Virgo represents the wheat harvest. He points out that the star ε Virginis is known as "the herald of the vintage," and the vintage comes considerably later in the year than the harvest. Aratos asserted that Leo first marked the harvest month, and this statement supports Maunder's argument.

According to the poets, this Virgin was Astræa, the daughter of Astræus and Aurora, and the goddess of justice. Near her appear the Scales in which, it is said, she weighed the good and evil deeds of men. In the golden age she resided in the earth, but becoming offended at the wickedness of mankind she returned to heaven. Hesiod claimed that she was the daughter of Jupiter and Themis, and Aratos gives more space to the history of this constellation in his celebrated poem than to any other constellation. His account is in part as follows:

Once on earth
She made abode, and deigned to dwell with mortals.
In those old times, never of men or dames
She shunned the converse; but sat with the rest
Immortal as she was. They call her Justice.
Gathering the elders in the public forum
Or in the open highway, earnestly
She chanted forth laws for the general weal,
Nor yet was known contention mischievous,
Nor fierce recrimination, nor uproar.
So lived they. Far off rolled the surly sea,
No ship yet from a distance brought supplies
But ploughs and oxen brought them. Queen of nations,
Justice herself poured all just gifts on man.
As long as earth still nursed a golden race
There walked she; but consorted with the silver
Rarely, and with reserves, nor always ready;
Demanding the old customs back again.

Nor yet that silver race she quite forsook.
At evening twilight, from the echoing mountains,
She came alone. No gracious words fell from her
But when the people filled the heights around
She threatened and rebuked their wickedness,
Refusing though besought to appear again;
"How have your golden fathers left a race
Degenerate! But you shall breed a worse
And then shall wars, and then shall hateful bloodshed
Be among men; and grief press hard on crime."
This said, she sought the mountains, and the people
Whose eyes still strained upon her, left for ever.
And when these also died, those others sprang,
A brazen race, more wicked than the last.
These first the sword, that roadside malefactor,
Forged; these first fed upon the ploughing oxen.
And Justice then, hating that generation,
Flew heavenward, and inhabited that spot
Where now at night may still be seen the virgin.

Virgo was also identified with Erigone, the daughter of Icarius, who hung herself when she learned of her father's death. In classic times she was associated with Ceres, or her daughter Proserpine. Proserpine, so the legend relates, was wandering in the fields in the springtime, and was carried off by Pluto to be his wife. Ceres besought Jupiter to intercede in the matter, and consequently Proserpine was allowed her liberty at intervals.

This myth is regarded as an allegory. Proserpine represents the seed which is buried in the earth, and in proper time bursts forth into bloom.

In Egypt Virgo was associated with Isis, and it was said that she formed the Milky Way by dropping innumerable wheat heads in the sky.

Another version of this myth is that Isis dropped a sheaf of corn as she fled to escape Typhon, which, as he continued to pursue her, became scattered over the heavens, thus producing the Galaxy which has all the appearance of glittering grains of golden corn.

The Chinese call the Milky Way "the Yellow Road,"

as resembling a path over which the ripened ears of corn are scattered.

The Egyptians represented Isis as holding three ears of corn in her hand. In the zodiacs of Denderah and Thebes the Virgin appears without wings, and holds in her hand an object said to be a distaff, marked by the stars in Coma Berenices.

In India Virgo was known as "the Maiden," and in the Cingalese zodiac she is represented as a woman in a ship with a stalk of wheat in her hand.

In the valley of the Euphrates the Virgin represented the goddess Istar, the daughter of Heaven, the Queen of the Stars. Istar was subsequently identified with Venus. The sign of the sixth month in the Akkadian calendar signified "the errand of Istar." According to Brown this errand was to seek her lost bridegroom in the underworld.

In China the Virgin was called "the Frigid Maiden," and the Chinese made the star group led by Spica the group of Spring.

The Arabs, who objected strongly to any drawing of the human figure, called Virgo "the Ears," because of the wheat ear that she held in hand, and Allen tells us that the early Arabs made from some members of the constellation the enormous Lion of the sky, and of others the Kennel Corner with Dogs barking at a Lion. Later Arabian astronomers referred to this constellation as "the Innocent Maiden."

Brown, in his *Stellar Theology* informs us that Virgo was identified as the goddess Rhea and adored under that name. This goddess was figured, according to Bryant, as a beautiful female adorned with a chaplet in which were seen rays composed of ears of corn (*i. e.*, wheat), her right hand resting on a stone pillar, and in her left hand appeared spikes of corn. By corn the ancients intended wheat. The spikes of "wheat" in the chaplet and left hand of the goddess Rhea are like those held in the left hand of

Photo by Brogi

Ceres
In the Vatican, Rome

Virgo, and emblematic of the season when the sun enters that sign.

Rhea was the daughter of sky and earth, the mother of Jupiter, and wife of Saturn, and also known as "Kronos" or "Time."

The association of Virgo with Rhea is of interest to Masons, as the goddess Rhea is the emblem of the Masonic Third Degree.

"Early Christian thought," says Maunder, "recognised a reference to the promise of the 'seed of the woman' of Genesis iii., 15, in 'the ear of corn' the Virgin carries in her hand, and the expression in Shakespeare's play of *Titus Andronicus* 'the good boy in the Virgin's lap,' refers to the mediæval representation of the sign as the Madonna and Child."

In the Hebrew zodiac Virgo is assigned to Napthali, whose standard was a tree, and in the land of Judæa Virgo was called "Bethulah."

Allen thinks that the custom of the Kern-Baby, that is still seen along the borderland of England and Scotland, was derived from the myths associated with Virgo, and that the tossing of the Corn Mother, a custom of La Vendée, was derived from a similar source.

Among the Peruvians Virgo was known as "the Magic Mother," and "the Earth Mother." The month festival was called "the Queen's festival," and was dedicated to the maize as well as to women in general, who in this month only predominated in the ritual.

Virgo has been associated with the Ashtoreth of the book of Kings, the Astarte of Syria, the Hathor of Egypt, and the Aphrodite of Greece.

In Assyria it was known as "Bel's Wife." In the Euphratean star list we find it styled "the Proclaimer of Rain."

Dr. Seiss identified this constellation with the Virgin Mary, and Cæsius associated Virgo with Ruth gleaning in the fields of Boaz. Schiller thought that the constellation represented James the Less, and Weigel regarded these stars as the Portuguese Towers.

The very ancient Sphinx of Egypt, the Riddle of the Ages, is thought by some to be a representation of Virgo's head on the body of Leo.

"Astrologically speaking," says Proctor, "Virgo is the joy of Mercury. Its natives (those born between the dates Aug. 22d and Sept. 23d), are of moderate stature, seldom handsome, slender but compact, thrifty and ingenious. It governs the abdomen, and reigns over Turkey, Greece, Mesopotamia, Crete, Jerusalem, Paris, Lyons, etc. It is a feminine sign and generally unfortunate. The cornflower is the significant flower and jasper the precious stone."

The constellation is noteworthy because of the great number of nebulæ found in this region of the heavens. The space embraced by the stars β, η, γ, δ Virginis, and Denebola in Leo, has been called "the Field of the Nebulæ." Sir William Herschel found here no less than 323 of these mysterious objects, which later search has increased to five hundred. This region of the sky was known to the Arabs as "the Kennel Corner of the Barking Dogs."

The beautiful white first magnitude star "Spica," α Virginis, is the most noted star in the constellation. It indicates the wheat ear which the Virgin holds in her left hand, and also signifies "the Ear of Wheat." The Arabs called it "the Solitary, the Defenceless, or Unarmed One," possibly because of its isolated position in the sky. They also knew it as "the Calf of the Lion," or "the Shin Bone of the Lion," Leo being much greater in extent in ancient times than is indicated by modern charts.

Spica forms with Denebola, Cor Caroli, and Arcturus the well-known figure of "the Diamond of Virgo."

Allen tells us that the Hindus knew this star as "Bright," figuring it as a Lamp, or Pearl, while the Chinese called it "the Horn" or "Spike." At one time in Egypt it was known as "the Lute Bearer."

In the Euphratean star list it bears the titles, "the Star

of Prosperity," "the Propitious One of Seed," "the One called Ear of Corn," and "the Corn Bearer."

Spica is especially interesting as furnishing Hipparchus the data which enabled him to discover the Precession of the Equinoxes. According to Lockyer, a temple at Thebes was oriented to Spica as early as 3200 B.C. Other temples oriented to this star are found at Olympia, Athens, and Ephesus. At Rhammus there are two temples almost touching each other, both following the shifting places of Spica. Many other temples were dedicated to Spica, and it seems to have been associated with the Min-worship of the Egyptians.

Spica is a spectroscopic binary, one of those stars which the spectroscope has shown to be attended by an invisible companion of enormous mass. Spica's dark companion revolves about it in a close orbit, making a complete revolution in the remarkably short period of four days. Spica is at such an enormous distance from us, that no reliable parallax has been obtained. Owing to its proximity to the ecliptic Spica is much used in navigation. It is a star of the Sirian type, and is said to be approaching our system at the rate of 9.2 miles a second. The star rises a very little south of the exact eastern point on the horizon, and culminates at 9 P.M., May 27th.

The star γ Virginis, known to the Latins as "Porrima," is an interesting star. Allen tells us that it is especially mentioned by Kazwini as being the "Angle" or "Corner of the Barker."

The Chinese knew it as "the High Minister of State."

It is a beautiful double star, and a fine sight in a small telescope, the two stars being about equal in brilliance, 3 and 3.2 magnitudes. In 1836 they showed as a single star, so close were they together, and consequently were indivisible even in the largest telescopes. Now they are 6″ apart, with a period of revolution estimated at about 190 years. Almost a complete revolution has been observed. The star culminates at 9 P.M., May 17th.

In the Alphonsine Tables ε Virginis was called "Vindemiatrix," signifying "Grape Gatherer," and the heliacal rising of this star was formerly the herald of the vintage time. The Arabs called it "the Forerunner of the Vintage." In a quatrain by Admiral Smyth we are told how to find this star:

Would you the star of Bacchus find on noble Virgo's wing,
A lengthy ray from Hydra's heart unto Arcturus bring;
Two thirds along that fancied line direct th' inquiring eye,
And there the jewel will be seen, south of Cor Caroli.

Photo by Bonfils

The Sphinx

The Galaxy
or Milky Way

THE GALAXY
OR MILKY WAY

The Milky Way: ah, fair illumined path,
That leadeth upward to the gate of heaven.

AMELIA.

WHO, of all those who have turned their eyes to the stars, has not wondered at that mysterious cloud that twines its devious way across the sky like a river's mist, awaiting the breath of the dawn-wind! And who, realising that this veil that flutters across the heaven is woven of a myriad close-set suns, has not felt a sense of awe and reverence steal upon him, a spirit of humility that takes possession of his soul!

The ancient Akkadians regarded the Milky Way as "a Great Serpent," or "the River of the Shepherd's Hut," and "the River of the Divine Lady."

Anaxagoras, who lived 550 B.C., and Aratos knew it as "το Γάλα," "that shining wheel, men call it milk."

The Greeks called it "the Circle of the Galaxy," and during all historic time it was regarded as "the River of Heaven," and "Eridanus," the Stream of Ocean.

In mythology it represented the stream into which Phaëton and the chariot of the sun were hurled by the enraged Jupiter.

Orientals fancied here a river of shining silver, whose fish were frightened by the new moon, which they imagined to be a hook.

Aside from the resemblance of the Galaxy to a serpent, and a river, the most popular notion of it among all people and in every age has been to regard the Milky Way as a highway amid the stars, the "Via Lactea" of the ancients. Chiefly it has been the road to heaven traversed by the souls of the departed.

The way to God's eternal house.

It may be interesting to review the peculiar ideas of the ancients respecting the Galaxy.

Anaxagoras thought that the Milky Way was a collection of stars whose light was partially obscured by the shadow of the earth.

Pythagoras said it was a vast assemblage of very distant stars.

Democritus about 460 B.C. held that the white cloudlike appearance of the Galaxy was due to the fact that, in that part of the heavens, there was a multitude of diminutive stars so close together that they illuminated each other.

It is strange that these early opinions of the Milky Way should have been confirmed in later days when the structure of this band of light could be examined in powerful telescopes, confirmed at least in the assumption that the white effect was produced by the presence of a myriad of stars.

Aristotle thought that the Galaxy was a vast mass of glowing vapour, far above the region of the ether and below that of the planets.

Parmenides believed that the milky colour was due to the mixture of dense and rare air.

Metrodorus and Oinopides conceived the strange idea that the Milky Way marked the pathway of the sun amid the stars.

Posidonius thought that it was a compound of fire less dense than that of the stars, but more luminous.

Theophrastus said it was the junction between the two hemispheres which together formed the vault of heaven, and that it was so badly made that some of the light supposed to exist behind the solid sky was visible through the cracks.

Plutarch claimed that the Galaxy was a nebulous circle which constantly appears on the sky, and, according to Blake, certain Pythagoreans asserted that when Phaëton lit the universe, one star which escaped from its proper place set light to the whole space it passed over in its circular course, and so formed the Milky Way.

Other philosophers imagined that the Galactic Circle was where the sun had been at the beginning of the world.

It was also believed that the Milky Way was but an optical phenomenon, produced by the reflection of the sun's rays from the vault of the sky as from a mirror, and comparable with the effects seen in the rainbow, and illuminated clouds.

Mythology attributed the Milky Way to the milk dropped from Juno's breasts. while she was suckling Hercules.

In Egypt Isis was said to have formed the Milky Way by the dropping of innumerable wheat heads.

There are many interesting legends concerning this celebrated pathway in the skies.

The ancients painted on the great canvas of the night skies many pastoral scenes, thus depicting features of daily life in the far east. Among the stars, as we have seen, we find the figure of a shepherd with his dogs watching his flocks, and near by twines a river, the Milky Way.

According to a French tale, the stars in the Milky Way are lights held by angel spirits to show mortals the way to heaven.

The Greeks called the Galaxy "the road to the Palace of Heaven." Along this road stand the palaces of the illustrious gods, while the common people of the skies live on either side of them.

The Algonquin Indians believed that this was the Path of Souls leading to the villages in the sun. As the spirits travel along the pathway their blazing camp-fires may be seen as bright stars.

Other Indian nations believed that the souls of the departed entered this pathway by the door situated where it intersects the zodiac in Gemini, and left it to return to the gods by the door of Sagittarius. The ancient Greeks and Romans had this same notion.

According to a Swedish legend, there once lived on earth two mortals who loved each other. When they died they were doomed to dwell on different stars far apart. They thought of bridging the distance between them by a bridge of light, and this bridge is to be seen in the Milky Way.

This tale is similar to the Japanese legend of the Milky Way, and the Star Lovers mentioned before. The Japanese call the Galaxy "the Silver River of Heaven," and believe that on the seventh day of the seventh month, the shepherd boy-star Altair and the Spinning Maiden, the star Vega, cross the Milky Way as on a bridge to meet each other. This happens only if the weather is clear, so that is why the Japanese hope for clear weather on the 7th of July, when the meeting of the Star Lovers is made a gala day throughout the kingdom.

The Danes regard the moon as a cheese formed by the milk that has run together out of the Milky Way.

In some parts of Germany Odin was considered identical with the Saxon god Irmin. Irmin was said to possess a ponderous chariot, in which he rode across the sky along the path which we know as the Milky Way, but which the ancient Germans called "Irmin's Way."

In the history of all nations and in all ages we find the Galaxy likened to a way, a road, or a pathway to the land of the hereafter.

Allen thinks that this universal idea may have come from the fancy that the heavenly way, crowded with stars, resembled the earthly road crowded with pilgrims.

The poets of all time have sung the praises of this bright pathway of the skies. Manilius thus refers to it:

A way there is in heaven's extended plain
Which when the skies are clear is seen below
And mortals by the name of milky know;

Photo by Anderson

The groundwork is of stars, through which the road
Lies open to the Thunderer's abode.

In Shakespeare's *Merchant of Venice* we read:

The floor of heaven
Is thick inlaid with patines of bright gold.
There 's not the smallest orb which thou behold'st
But in his motion like an angel sings,
Still quiring to the young-eyed Cherubim.

Sir John Suckling says:

Her face is like the Milky Way i' the sky,
A meeting of gentle lights without a name.

In Milton's *Paradise Lost* there is this beautiful reference to the Galaxy:

A broad and ample road whose dust is gold
And pavement stars as stars to thee appear
Seen in the Galaxy, that Milky Way
Which nightly as a circling zone thou seest
Powdered with stars.

Longfellow thus alludes to the Milky Way in *Hiawatha:*

Showed the broad white road in heaven,
Pathway of the ghosts, the shadows.
Running straight across the heavens,
Crowded with the ghosts, the shadows,
To the kingdom of Ponemah,
To the land of the hereafter.

When Galileo directed his newly invented telescope at the Galaxy the mystery of its composition was solved. Myriads of stars strewed the fields as he swept over the misty belt, their blended light causing the white effect the unaided eye reveals.

Allen thus sums up our present knowledge of this remarkable object:

"It covers more than one tenth of the visible heavens, containing more than nine tenths of the visible stars, and seems a vast zone-shaped nebula nearly a great circle of the sphere, the poles being in Coma and Cetus."

The Milky Way has been likened to sand strewn not evenly as with a sieve, but as if flung down by handfuls and both hands at once, leaving dark intervals, and all consisting of stars of the 14th, 16th, and 20th magnitudes down to nebulosity.

It is believed that the majority of stars comprising this wonderful belt of stars surpass our sun in brilliancy and splendour. In the deep recesses of this glittering way Sir Wm. Herschel was able to count five hundred stars receding in regular order behind each other, and in the interval of an hour 116,000 stars passed him in review across the field of his telescopic vision.

In the constellation Cygnus, where the Milky Way is especially brilliant, there is a region about five degrees in breadth which contains, it is said, 331,000 stars.

Prof. Russell writing of this region says: "Here the Milky Way is crossed by a dark streak which immediately suggests a passing cloud. But, year in and year out, on the clearest nights, the dark region is there. Its origin must be interstellar space—perhaps in an actual thinning of the stars of the Galaxy, perhaps in the interposition of some cosmic cloud of overwhelming vast dimensions."

Many think the Galaxy a universe by itself and our sun one of its myriad stars.

"It remains the most wonderful sight that human eyes behold. The thought of its wonderful structure, the contemplation of the splendour proximity would afford, transcends the very limits of the human intellect, and gives us a mere glimpse in imagination of the stupendous scale of a universe of which our system is but an infinitesimal atom."

The following are some of the titles bestowed on the Milky Way, and various fancies concerning it:

The Akkadians imagined it to be a Great Serpent, and the River of the Divine Lady.

The Greeks called it "the Circle of the Galaxy."

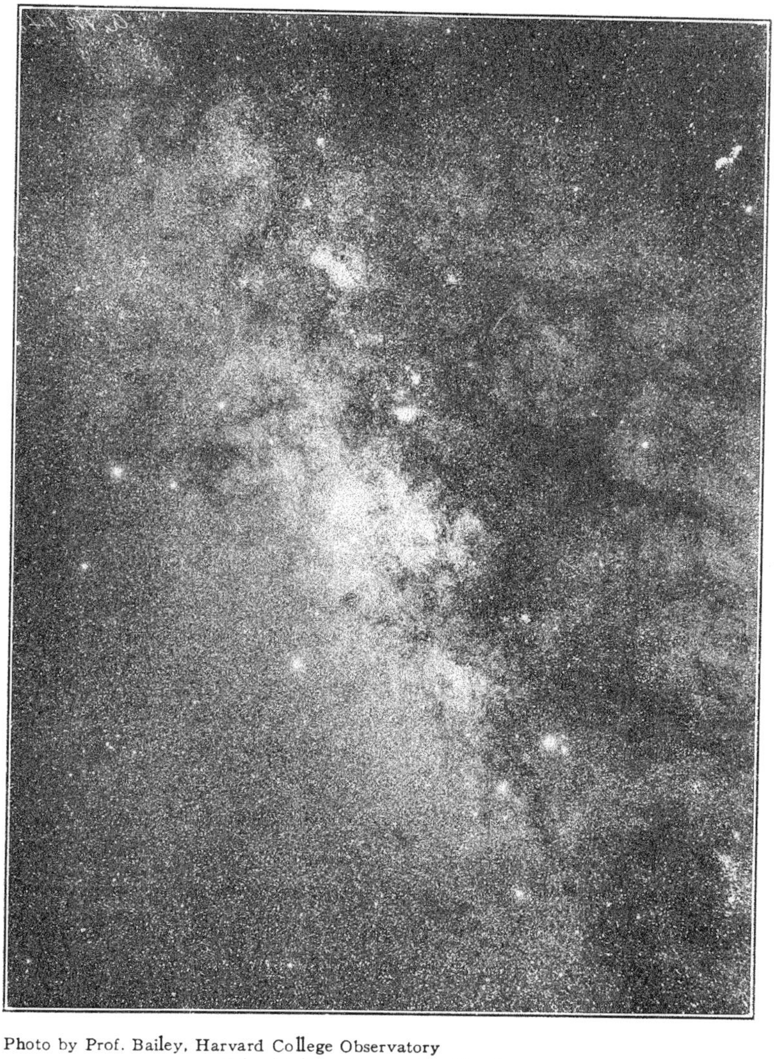

Photo by Prof. Bailey, Harvard College Observatory

The Milky Way in Sagittarius

In Rome it was regarded as "the Heavenly Girdle," and as a Circle.

The ancient inhabitants of Britain called it "Watling Street."

One of the Celtic titles is "King of Fairies," and the Celts also fancied that it was the road along which Gwydyon pursued his erring wife.

In the mediæval ages it was known as "the Way to Rome."

The ancient Germans called it "Irmin's Way"; Germans to-day call it "Jacob's Road," while the French peasants call it "St. James's Road."

The Norsemen and Scandinavians knew it as the path to Valhalla, up which went the souls of heroes who fell in battle. The Swedish peasantry call it "Winter Street."

In Japan and China it was known as "the Celestial River," and "the Silver River." The Chinese also called it "the Yellow Road."

The Arabs knew it as "the River," while the Eskimos of the far north call it "the Path of White Ashes." The Bushmen, far removed from these dwellers in the Frigid Zone, thought that the Milky Way was composed of wood ashes thrown up into the sky by a girl, that people might see their way home at night.

The Australians call it "the fire smoke of an ancient race." The Masai name for it is "the road across the sky."

The Dutch, Basutos, and Zulus call it "the neck of the sky."

The Peruvians and the Incas knew it as the "dust of stars," while the Patagonians thought that it was the road on which their dead friends were hunting ostriches.

The early Hindus knew it as "the Path of Āryamān" leading to his throne in Elysium.

In the Punjab it was "the Path of Noah's Ark," while in northern India it was "the Path of the Snake."

The Ottawa Indians believed it to be the muddy water

stirred up by a turtle swimming along the bottom of the sky.

The North American Indians regarded the Galaxy as the pathway of the ghosts to the land of the hereafter, the Pawnees believing that it was the path taken by spirits as they pass along, driven by the wind which starts at the north to the star in the south at the end of the way. The Iroquois call it "the Road of Souls."

The Tahitians regard it as a shark-infested creek, and the Polynesians knew it as the "Long Blue Cloud-eating Shark."

It has also been called "Walsingham's Way," "the Road to the Virgin Mary in Heaven," "Asgard's Bridge," and "the Band."

The Hyades and Pleiades

The Hyades

THE HYADES AND PLEIADES

THE HYADES

Who hears not of the Hyades, sprinkling his forehead o'er?

ARATOS.

THE "V"-shaped group of stars in the constellation Taurus is known as "the Hyades," and has attracted the attention of all mankind. Its stars outline the face of the fierce Bull that lowers its massive head to gore the giant hunter Orion, and the ruddy first magnitude star Aldebaran, the lucidă of the group, marks the eye of the enraged creature.

According to Allen the Greeks knew this star cluster as Ὑάδες, which became Hyades with the cultured Latins, a title supposed by some to be derived from ὕειν, "to rain," referring to the wet period attending their morning and evening setting in the latter parts of May and November, and this is their universal character in the literature of all ages.

The poets call these stars the "rainy Hyades" or the "watery Hyades." Thus Horace in his ode to the ship bearing Virgil to Greece sings:

In oak or triple brass his breast was mailed
Who first committed to the ruthless deep
His fragile skiff . . .
Nor feared to face the tristful Hyades.

Manilius refers to them as "sad companions of the turning year."

Pliny called them collectively "a violent and troublesome star, causing storms and tempests raging both on land and sea."

26

Spenser thus describes their setting:

> And the moist daughters of huge Atlas strove
> Into the ocean deepe to drive their weary drove.

Virgil alludes to them as "the rainy Hyades," and Tennyson in his *Ulysses* wrote:

> Thro' scudding drifts the rainy Hyades vext the dim sea.

So we find them treated consistently, and always identified with a rainy period of weather.

The Romans thought that the Greek name Hyades was derived from ὗες, meaning sows, so they called these stars "suculæ" or little sows, and owing to this error much confusion has arisen. Pliny accounts for the title by the fact that the continued rains of the season of the setting of the Hyades made the roads so miry that these stars seemed to delight in dirt like swine. This explanation, however, seems far fetched.

Isidorus claimed that the title "suculæ" was derived from "sucus," meaning moisture, which idea fits in very well with the watery traditions that have always surrounded this group of stars.

Some authorities derived the name of this group from the letter "Y," to which its form, bears a resemblance, though the stars are grouped more in the shape of the letter "V."

The Hyades have also been called "a Torch," "a Triangular Spoon," and "the Little She Camels," the large camel being represented by the star Aldebaran.

The Hindus saw here a temple or waggon, the Chinese, a hand-net, or rabbit-net, but the latter generally called the group "the Star of the Hunter" or "the Announcer of Invasion on the Border." They worshipped these stars as "the General or Ruler of Rain," from at least 1100 B.C.

According to Grimm the Hyades were regarded as "the Boar Throng" among the Anglo-Saxons.

In mythology the group were supposed to be the daughters of Atlas, and half-sisters of the Pleiades. They were changed into stars on account of their grief for the death of their brother Hyas.

According to another story they were the nurses of the infant Bacchus, and the father of the gods rewarded them for their faithful service by placing them among the stars. Originally they were supposed to be seven in number. Hesiod named five, and we now regard the group as containing six stars. The Hyades are among the few stellar objects mentioned by Homer.

Aldebaran, or Alpha Tauri, the lucida of the group, rises an hour later and almost directly under the celebrated star cluster known as the Pleiades, and its name indicates the fact, meaning "the hindmost" or "the follower."

Mrs. Martin, in *The Friendly Stars*, gives us the following facts concerning this beautiful star: "Aldebaran is the fourteenth star in order of brightness in the entire heavens, and the ninth among those seen in our latitude. It is what is known as a standard first magnitude star. It gives us about one ninety billionth as much light as the sun, but at the same distance as the sun we would get from it forty-five times as much light as the sun gives us. It requires something more than thirty-two years for the light of Aldebaran to reach the earth, which means that it is nearly two hundred trillions of miles away. It is increasing this distance at the rate of about thirty miles a second but even at this rate it will require more than ten thousand years to add another trillion of miles to its distance."

Its red hue indicates that Aldebaran is one of the older stars, one of the suns that like a dying ember still glows persistently as if in anger at the loss of its pristine glory and the thought of its declining power.

According to Prof. Russell the Hyades are receding from us at the rate of twenty-five miles a second, and on the average its stars are about one hundred and twenty light years distant. At this distance our sun would appear as a telescopic eighth magnitude star.

The group is exceedingly rich in double stars, and viewed even in a small telescope with a low power presents a beautiful appearance.

Serviss thus mentions the group: "The beauty of Aldebaran, the singularity of the figure shaped by its attendants, the charming effect produced by the flocks of little stars, the Deltas and the Thetas, in the middle of the arms of the letter, and the richness of the stellar groundwork of the cluster, all combine to make the Hyades one of the most memorable objects in the sky; but no one can describe it, because the starry heavens cannot be put into words."

The Pleiades

THE PLEIADES

> Open those Pleiad eyes, liquid and tender,
> And let me lose myself among their depths.
>
> DE VERE.

No group of stars known to astronomy has excited such universal attention as the little cluster of faint stars we know as "the Pleiades." In all ages of the world's history they have been admired and critically observed. Great temples have been reared in their honour. Mighty nations have worshipped them, and people far removed from each other have been guided in their agricultural and commercial affairs by the rising and setting of these six close-set stars.

Mrs. Martin thus charmingly alludes to them:

"The magic of their quivering misty light has always made a strong appeal to men of imagination. Minstrels and poets of the early days sang of their bewitchment and beauty, and many of the great poets from Homer and the author of Job down to Tennyson and the men of our own day have had their fancy livened by them and in one form or another have celebrated their sweetness and mystery and charm."

Many have been the metaphors inspired by this famous cluster. They have been compared to a rosette of diamonds, to a swarm of fireflies or bees, and the shining drops of dew. More prosaic minds have regarded these stars as a hen surrounded by her chickens, and some have thought that they represented the seven virgins.

"Even with people who do not know them by sight and have not felt the sweet influences of the Pleiades, there is a vague memory of some story about a lost Pleiad that stirs an emotion suggesting something romantic and sad. The Pleiades form in truth a delightful group of twinkling unfathomable stars, singularly fascinating and singularly persistent in their brilliancy."—Mrs. Martin.

On the Euphrates the Pleiades and the Hyades were known as "the Great Twins of the Ecliptic." The Babylonians and Assyrians regarded them as a family group without dreaming of the full significance of the title, for modern science has proved that this group of suns have a common proper motion, that is, they are moving through space in the same direction, and are obviously part of one great system that holds them fast in bonds immutable.

The patriarch Job is thought to refer to the Pleiades in his word "Kimah," meaning "a cluster or heap," which occurs in the Biblical passages: "[God] maketh Arcturus, Orion, and the Pleiades and the Chambers of the South," and the familiar query: "Canst thou bind the sweet influences of the Pleiades or loose the bands of Orion?"

The meaning of this inquiry has been the cause of much conjecture and many attempts have been made to interpret the sense of it. Maunder thus explains the passage: When the constellations were first designed the Pleiades rose heliacally at the beginning of April and were the sign of the return of spring. Aratos wrote of them:

> Men mark their rising with the solar rays,
> The harbinger of Summer's brighter days.

The Pleiades which thus heralded the return of this genial season were poetically taken as representing the power and influence of spring. Their "sweet influences" were those that rolled away the gravestone of snow and ice which had lain upon the winter tomb of nature.

The question of Job was in effect, "What control hast thou over the powers of nature? This is God's work, what canst thou do to hinder it?" Of the sweet influence of these fair stars we read again in Milton's *Paradise Lost*, where the poet sings of the Pleiades in the morning skies:

The Dance of the Pleiades
Painting by Elihu Vedder

> . . . the grey
> Dawn and the Pleiades before him danced,
> Shedding sweet influence."

In the New Testament we find the "Seven Stars" also mentioned. In the first chapter of the Revelation, the Apostle St. John writes that "he saw seven golden candlesticks and in the midst of the seven candlesticks one like unto the Son of Man . .. . and He had in his right hand Seven Stars. The Seven Stars are of the angels of the seven churches, and the seven candlesticks are the seven churches."

The Seven Stars in a simple compact cluster, says Maunder, stand for the church in its many diversities, and its essential unity. Modern almanacs designate the Pleiades "the 7*" or "seven stars."

The Pleiades were among the first mentioned stars in the astronomical literature of China, one record of them bearing the early date of 2357 B.C., when Alcyone, the lucida of the group, was near the vernal equinox. The Chinese young women worshipped these stars as the Seven Sisters of Industry.

As might be expected, this celebrated group was the object of worship in Egypt. There the Pleiades were identified with the goddess Nit, meaning the shuttle, one of the principal divinities of Lower Egypt.

The Great Pyramid, which was without doubt erected for astronomical purposes, is closely associated with the Pleiades, as Proctor has shown.

In the year 2170 B.C. the date at which the Pleiades really opened the spring season by their midnight culmination, they could be seen through the south passageway of this gigantic mausoleum. It has even been suggested that the seven chambers of the Great Pyramid commemorate these seven famous stars. Blake says: "Either the whole of the conclusions respecting the pyramids is founded on pure imagination, or we have here another remarkable proof of the influence of the Pleiades on the reckoning of the year."

The Egyptians called this star group "Athur-ai" or "Atauria," meaning the stars of Athyr (Hathor), a name also given the Seven Stars by the Chaldeans and Hebrews. From this title is derived the Latin Taurus, and the German Thier. It is possible that this title was somehow connected with the Greek letter tau, and the sacred scarabæus or tau beetle of Egypt. It has also been suggested that the "tors" and Arthur's Seat, which were names given to British hilltops, may be connected with the "high places" of the worship of the Pleiades. Arthur's Seat at Edinburgh is a notable example of such a site.

The Arabians called the Pleiades, "Atauria" signifying "the little ones."

There appear to be three distinct derivations of the word Pleiades. First, from the Greek word πλεῖν, meaning "to sail," the heliacal rising and setting of these stars marking the opening and closing of the season of navigation among the Greeks.

Second, from πέλειαι, meaning "a flight of doves." Hesiod, Pindar, and Simonides all use this word. The doves or pigeons were considered as flying from the mighty hunter Orion. They were also said to be the doves that carried ambrosia to the infant Zeus.

D'Arcy Thompson asserts that the Pleiad is in many languages associated with bird names, and considers that the bird on the bull's back on coins of Eretria and Dicæa represents the Pleiades. We have a reduplication of this strange position of a bird among the constellational figures in the crow perched on the coils of Hydra.

A third derivation of the title of this group is from πλεῖος, meaning "full" or in the plural "many." This derivation is considered to be the correct one by the weight of authority.

Many of the Greek temples were oriented to the Seven Stars, notably temples erected as early as 1530 and 1150

B.C., and the noted Parthenon built in 438 B.C., and in the works of the Grecian poets we find many references to the group.

Allen tells us that the Hindus pictured these stars as a flame typical of Agni, the god of fire, and regent of the asterism. The more usual representation of the group among the Hindus was a razor; possibly the arrangement of the stars in the group suggested this title. It is thought that there may be a connection between the Hindu title "Flame," and the great Feast of Lamps of the western Hindus held in the Pleiad season, October and November, a great festival of the dead which gave rise to the present Feast of Lanterns of Japan.

This closely associated star group has from time immemorial impressed mankind with a sense of mystery. A great cataclysm, possibly the Biblical Deluge, is in some way connected with the Pleiades, and some reference to such an event can be traced in many of the legends and myths surrounding these stars that have come down to us from nations far removed from each other.

Memorial services to the dead at the season of the year when the Pleiades occupied a conspicuous position in the heavens are found to have taken place, and to have been a feature in the history of almost every nation of the earth, from remote antiquity to the present day. The universality of this custom may well be considered one of the most remarkable facts that astronomical history records, and it serves to make the study of this group the most interesting chapter in all stellar history. This little group of stars, twinkling so timidly in the nights of autumn in the eastern heavens, links the races of mankind in closer relationship than any bonds save nature's. No wonder that they have inspired universal awe and admiration, that within this group of suns man has sought to find the very centre of the universe.

Among the Aztecs of South America we find the Pleiades the cynosure of all eyes, a nation trembling at their feet.

At the end of every period of fifty-two years, in the month of November when the Pleiades would culminate at midnight, these rude people imagined the world would end. Human sacrifices were offered, while the entire population passed the night upon their knees awaiting their doom.

Far removed from the Aztecs we find the people of Japan in their great national festival, the Feast of Lanterns, a feast that is alive to-day, commemorating at this same season of the year some great calamity which was supposed to have overwhelmed the race of man, in the far distant past, when these seven little stars were prominent in the heavens.

In the Talmud we find a curious legend associating the Pleiades with an all-destroying flood, expressed as follows:

"When the Holy One, blessed be He, wished to bring the deluge upon the world, He took two stars out of the Pleiades and thus let the deluge loose, and when He wished to arrest it, He took two stars out of Arcturus and stopped it."

As we have seen, the ancient Hindus, the Aztecs, and the Japanese all had memorial festivals in the month of November. These generally occurred on the 17th of the month.

Among the ancient Egyptians the same day was observed, and although their calendar was subsequently changed, the occasion was not lost sight of. The date of their celebration was determined by the culmination of the Pleiades at midnight, and on this date the solemn three days' festival commenced. With them, as with the three previously mentioned nations, the festival was associated with the tradition of a deluge or race-destroying calamity. Blake says in regard to this that "when we connect the fact that this festival occurred on the 17th day of Athyr, with the date on which the Mosaic account of the deluge of Noah states it to have commenced, in the second month of the Jewish year, which corresponds to November, the 17th day of the month, it must be acknowledged that this is no chance coincidence, and that the precise date here stated must have been regulated by the Pleiades, as was the Egyptian date." Surely this is an interesting reference to the history of these stars.

The Persians formerly called the month of November "Mordad," meaning "the angel of death," and that month marked the date of their festival of the dead. On the day of the midnight culmination of the Pleiades, Nov. 17th, no petition was presented in vain to their ancient kings.

In Ceylon, and in far distant Peru, a like festival took place at this season of the year. In the latter country the observation of the rising and setting of the Pleiades was the basis of their primitive calendar.

The Society Islanders commenced their year on the first day of the appearance of the Pleiades, which occurred in November. This star group also marked a festival in commemoration of the dead which took place annually about the end of October in the Tonga Islands of the Fiji group.

Blake tells us that the first of November was with the ancient Druids of Britain a night full of mystery, in which they annually celebrated the reconstruction of the world. Although Druidism is now extinct the relics of it remain to this day, for in our calendar we still find Nov. 1st marked as "All Saints' Day," and in the pre-Reformation calendar the last day of October was marked "All Hallow Eve," and the 2d of November as "All Souls'," indicating clearly a three days' festival of the dead, commencing in the evening, and originally regulated by the Pleiades.

In France, the Parisians at this festival repair to the cemeteries and lunch at the graves of their ancestors. Prescott in his *History of the Conquest of Mexico*, states that the great festival of the Mexican cycle was held in November at the time of the midnight culmination of the Pleiades, and the Spanish conquerors found in Mexico a tradition that the world was once destroyed when the Pleiades culminated at midnight, the identical tradition that we find in the far east, a myth so universal as to suggest a foundation of fact.

The actual observance at the present day of this festival is to be found among the Australian savages. At the midnight culmination of the Pleiades, in November, they still hold a New Year's corroboree in honour of this group of stars, which they say are "very good to the black fellows." The corroborees are connected with a worship of the dead. Still another custom associated with the Pleiades which has come down to us is the November date of our elections; the convocation of the tribal meeting at this time, because of the significant position of the Pleiades, being a very ancient custom.

Many Masonic organisations of the present day have memorial services to the dead about the middle of November, a survival of the universal recognition of the season of the year as commemorating the destruction of the world, when the Pleiades culminated at midnight.

The fall of the year was especially appropriate as a season for memorial services for the dead, as nature's life was then at a low ebb and every prospect was suggestive of death, and the preparation for the long sleep imposed by winter. Thus we see in the association of this star group with this season of the year, a link that binds the remote past with the ever-living present in a most remarkable manner, and no one cognisant of these facts can watch these faintly glimmering stars with any feelings save those of awe and reverence.

Brown tells us that in the symbolism of Masonry the Pleiades play a prominent part. The emblem of the Seven Stars alludes to this star group as emblematic of the vernal equinox, thus making the Pleiades a beautiful symbol of immortality. It was for this reason that of all the "hosts of heaven" the Pleiades were selected as an emblem.

In ancient times the appearance and disappearance of the Pleiades was associated with meteorological conditions. Statius calls them "a snowy constellation." Valerius

Flaccus speaks of their danger to ships, and Horace pictures the south wind lashing the deep into storm in the presence of these famous stars. The Romans generally referred to the Pleiades as "Vergiliæ" or "Virgins of Spring."

This star cluster was also of great service to the husbandman in marking the progress of the year. Hesiod thus alludes to the Pleiades:

> There is a time when forty days they lie
> And forty nights concealed from human eye,
> But in the course of the revolving year,
> When the swain sharps the scythe, again appear.

He also refers to the rising of the Pleiades as the time for the harvest, while the period at which they disappeared for some time, he termed ploughing time.

The heliacal rising of this star group, that is its rising with the sun, heralded the summer season, while its acronical rising, when it rose as the sun set, marked the beginning of winter, and led to the association of the group with the rainy season, and with floods, so often mentioned by the poets. Aratos thus expressed its acronical rising:

> Men mark their rising with Sol's setting light,
> Forerunners of the Winter's gloomy night.

Valerius Flaccus used the word "Pliada" for showers, and Josephus tells us that during the siege of Jerusalem by Antiochus Epiphanes, in 170 B.C., the besieged wanted for water until relieved by a large shower of rain which fell at the setting of the Pleiades.

Pope in his "Spring" thus alludes to the showery nature of the Pleiades:

> For see: the gath'ring flocks to shelter tend,
> And from the Pleiades fruitful showers descend.

Among the Dyaks of Borneo, the Pleiades regulated the seasons by their periodic return and disappearance, and guided them in their agricultural pursuits.

In South Africa, they were called the "hoeing stars," and their last visible rising after sunset has been celebrated with rejoicing all over the southern hemisphere as betokening the summons to agricultural activity.

The Bantu tribe called the group "the ploughing constellation," because its rising in the early morning in midwinter told the black man to turn out in the cold and plough for mealies. With the Peruvians also the Pleiades governed the crops and harvest, and indeed were supposed to have created them.

Four thousand years ago this star group marked the position of the sun at the spring equinox, and this is the principal reason why, as we have seen, it was so universally associated with the apparent wax and wane of the forces of nature.

Many strange fables and fancies surround the Pleiades quite apart and entirely disassociated with their classical mythology. The Hottentots had a curious notion concerning them. They regarded the Pleiades as wives who shut their husbands out because they missed their game. It would be difficult to trace the origin of this singular idea concerning these stars.

The Pleiades was the favourite constellation of the Iroquois Indians. In all their religious festivals the calumet was presented towards these stars, and prayers for happiness were addressed to them. They also believed that the Pleiades represented seven young persons who guarded the holy seed during the night.

An Onondaga legend concerning these stars is as follows: "A long time ago a party of Indians journeyed through the woods in search of a good hunting ground. Having found one, they proceeded to build their lodges for the winter, while the children gathered together to dance and sing. While the children were thus engaged, an old man dressed in white feathers, whose white hair shone like silver, appeared among them and bid them cease dancing lest evil befall them, but the children danced on unmindful of the warning, and presently they observed that they were rising little by little into the air, and one exclaimed, 'Do not look back for something strange is taking place.' One of the children disobeyed this warning and looking back became a falling star. The other children reached the high heavens safely and now we see them in the star group known as the Pleiades."

Another Indian legend relates that "seven brothers once upon a time took the warpath and discovered a beautiful maiden living all alone whom they adopted as their sister. One day they all went hunting save the youngest, who was left to guard his sister. Shortly after the departure of the hunters, the younger brother discovered game and set off in pursuit of it, leaving his sister unprotected. Whereupon a powerful buffalo came to her lodge and carried her away. The brothers returned and in dismay found that their sister had been taken from them. They immediately went in pursuit of her, only to find that she was confined in a lodge in the very centre of a great herd of fierce buffaloes. The younger brother cleverly tunnelled beneath them, however, and rescued his sister, and hastened homeward with her, where her brothers hedged her lodge about with a very high iron fence. The buffaloes, enraged at the escape of the maiden, attacked the seven brothers, and battered down the fence, only to find that the maiden and her brothers had been carried upward to the sky out of their reach, and there they may be seen in the clustering Pleiades."

The Shasta Indians of Oregon have the following legend concerning the Pleiades:

"The Coyote went to a dance with the Coon. On his return home he sent his children after the game he had killed, and when they had brought it in, he prepared a grand feast. The youngest child was left out, and in anger went to the Coon's children and told them that the Coyote had killed their father. The Coon's children revenged themselves by killing all the Coyote's children, save one, while the Coyote was away from home. They then disappeared.

The Coyote, being unable to find his children, hunted everywhere, and asked all things as to their whereabouts. As he was searching he perceived a cloud of dust rising, and in the midst he saw the Coon's children and his youngest child. He ran after them in vain, and the children rose to the stars where they became the Pleiades." The Coyote's child is represented by the faintest star of the group.

In winter, when Coons are in their holes, the Pleiades are most brilliant, and continually visible. In summer, when Coons are out and about, the Pleiades are not to be seen.

The medicine men among the Malays, in their invocations, besought the Pleiades to help them heal bodily diseases. The Abipones, a tribe of Indians dwelling on the banks of the Paraguay River in South America, thought that they were descended from the Pleiades, and as that asterism disappeared at certain periods from the sky of South America, upon such occasions they supposed that their grandfather was sick, and were under a yearly apprehension that he was going to die, but as soon as the seven stars were again visible in the month of May, they welcomed their grandfather as if restored from sickness with joyful shouts and the festive sound of pipes and trumpets, and congratulated him on the recovery of his health. The hymn of welcome begins: "What thanks do we owe thee? And art thou returned at last? Ah! thou hast happily recovered."

Maunder tells us that in many Babylonian cylinder seals there are engraved seven small discs in addition to other astronomical symbols. These seven discs are arranged thus:

. . . .   - - -
or
. . .   . . .

much as we would plot the Pleiades. In all probability these discs represent this celebrated star group.

Another name for the Pleiades was "the clusterers," and they are frequently represented on ancient coins by a cluster of grapes. A coin of Mallos in Cilicia shows them represented by doves whose bodies are formed by bunches of grapes.

The Pleiades according to mythology were the seven daughters of Atlas, the giant who bears the world upon his shoulders, and the nymph Pleione. The story is, that these seven maidens, together with their sisters the Hyades, were transformed into stars on account of their "amiable virtues and mutual affection." According to Æschylus they were placed in the heavens on account of their filial sorrow at the burden imposed upon their father Atlas.

Aratos thus records the names of these seven sisters:

> These the seven names they bear:
> Alcyone and Merope, Celæno,
> Taygeta, and Sterope, Electra,
> And queenly Maia, small alike and faint,
> But by the will of Jove illustrious all
> At morn and evening, since he makes them mark
> Summer and winter, harvesting and seed time.

One myth concerning the Pleiades relates that they were so beautiful in appearance that Orion unceasingly pursued them, much to their discomfiture. They appealed to Jupiter for assistance and he pitying them changed them into doves. Thereupon they flew into the sky and found a refuge among the stars.

The Smith Sound Eskimos have the following legend concerning the Pleiades, which group they call "Nanuq," meaning "the Bear": "A number of dogs were pursuing a bear on the ice. The bear gradually rose up in the air as did the dogs until they reached the sky. Then they turned to stars and the bear became a larger star in the centre of the group, and is represented by the star Alcyone."

One of the seven stars in this cluster is not as brilliant as the others and this star the Greeks called "the Lost Pleiad."

The tradition that one of the stars of this group has been lost or has grown dim is very ancient and almost universal. It is found among nations far removed from each other and has survived to the present day. It is found in Greece, Italy, and Australia, among the Malays in Borneo, and the negroes of the Gold Coast.

Miss Clerke writes: "Variants of the classical story of the 'Lost Pleiad' are still repeated by sable legend-mongers in Victoria, by head-hunters in Borneo, by fetish worshippers amid the mangrove swamps of the Gold Coast. An impression thus widely diffused must either have spread from a common source or originated in an obvious fact; and it is at least possible that the veiled face of the seventh Atlantid may typify a real loss of light in a prehistorically conspicuous star."

Byron thus alludes to this mysterious star:

> Like the lost Pleiad seen no more below;

and Aratos wrote:

> As seven their fame is on the tongues of men,
> Though six alone are beaming on the eye.

There is little doubt that originally one of these stars was brighter than it now appears. Some of the Pleiades are known to be variable, and one of them may have lost lustre at some time far remote, a fact that may account for the tradition of a lost star.

It is interesting to review the myths and legends of the Lost Pleiad and the ingenious suggestions that have been made to account for its apparent loss of brilliancy.

As to which of the seven sisters disappeared mythology is uncertain. According to one story it was Electra, the mother of Dardanus, the founder of Troy, who hid her face in order that she might not see the destruction of that city. The Greeks claimed that the Lost Pleiad was Merope, who marrying a mortal, and feeling disgraced, withdrew from the company of her sisters. Some said the seventh Pleiad

The Lost Pleiad
By Randolph Rogers

was struck by lightning, others that it was removed into the tail of the Great Bear. There is a myth that while a terrible battle was being waged on the earth, one of the sisters hid herself behind the others. The Iroquois Indians also had a legend respecting this famous star that appears to have been lost. They imagined that the Lost Pleiad was a little Indian boy in the sky, who was very homesick. When he cried he covered his face with his hands and thus hid his light. The legend is as follows: "Seven little Indian boys lived in a log cabin in the woods, and every starlight night they joined hands and danced about singing the 'Song of the Stars.' The stars looked down and learned to love the children, and often beckoned to them. One night the children were very much disappointed with their supper, and so when they danced together and the stars beckoned to them, they accepted the invitation and betook themselves to Starland, and became the seven Pleiades, and the dim one represents one of the little Indian boys who became homesick."

According to another legend concerning the Lost Pleiad, known to be current among the blacks of Australia, this star group represented a queen and her six attendants. Long ago the Crow (our Canopus) fell in love with the queen, who refused to be his wife. The Crow found that the queen and her attendants were wont to hunt for white edible grubs in the bark of trees, and changing himself into a grub hid beneath the bark. The six maidens sought in vain to pick him out with their wooden hooks, but when the queen tried to draw him out with a pretty bone hook he came out, and assuming the shape of a giant ran away with her. Ever since that time there have been only six stars in the group.

Aratos wrote of the number of the Pleiades:

> Seven paths aloft men say they take,
> Yet six alone are viewed by mortal eyes.
> From Zeus' abode no star unknown is lost
> Since first from birth we heard, but thus the tale is told.

Euripides mentions these "seven paths," and Eratosthenes calls them "the seven-starred Pleiad," although he describes one as "all invisible."

The South Sea Islanders' myth concerning the Pleiades relates that these stars were once a single star which shone with such a clear lustre as to incur the envy of the god Tane, who was in league with the stars Aldebaran and Sirius and followed the Pleiades. Tane in his anger hurled Aldebaran at this bright star and broke it up into six parts, each of which became a star.

The blacks of Victoria, Australia, have a myth in which the Pleiades are considered a host of young wives. Another myth relates that these stars were once pretty maidens on the earth who were followed by some young men called "the Beriberi." To get away from them, the girls climbed into the tree-tops, and thence sprang into the heavens, where they were transformed into shining bodies. One maiden remained behind. She was called "the shy one," and is represented by the least bright star in the group. The Beriberi were eventually placed in the heavens where they appear in the girdle of Orion.

In the Solomon Islands the Pleiades were also called a company of maidens.

The Dyaks and the Malays of Borneo imagine the Pleiades to be six chickens followed by their mother, who remains always invisible. At one time there were seven chickens, but one of them paid a visit to the earth, and there received something to eat. This made the hen very angry and she threatened to destroy the chickens, and the people on the earth. Fortunately the latter were saved by Orion, the mighty hunter. At that period of the year when the Pleiades are invisible the Dyaks say that "the hen broods her chickens." When these stars are to be seen they say "the cuckoo calls."

The North American Indians call the Pleiades "the dancers," while the South American Indian name for the group is "the six stars." The cluster has also been likened to a necklace of brilliant gems, and popularly associated with the "little she goats" that Sancho Panza saw on his aërial excursions.

We come now to a consideration of the individual stars of this celebrated group. Alcyone, the brightest of the Pleiades, represents in the sky the Atlantid nymph who became the mother of Hyrieus by Poseidon. It is sometimes called "the light of the Pleiades." The Arabs called it "the bright one" and "the Walnut." Alcyone is famous as locating the supposed centre of the universe, the point about which the starry heavens revolved. The German astronomer Mädler held this view, but there is no satisfactory reason for his opinion. Alcyone has three companion stars, and the three form a beautiful little triangle, a fine sight in a small telescope. The star culminates at 9 P.M., on the last night of the year. Miss Clerke considers that Alcyone exceeds the sun in brilliancy one thousand times.

Beyond the moons that beam, the suns that blaze,
Past fields of ether, crimson, violet, rose,
The vast star-garden of eternity,
Behold: it shines with white, immaculate rays,
The home of peace, the haven of repose,
The lotus-flower of heaven, Alcyone.

Frances Mace.

The star Maia represents the oldest and most beautiful of the sisters, and some have said that this star was the brightest of the group. Maia married Jupiter and became the mother of Mercury, of whom Shelley sings:

Farewell, delightful boy,
Of Jove and Maia sprung—never by me
Nor thou, nor other songs, shall unremembered be.

It was discovered in 1884 that Maia was surrounded by a nebulous cloud, while later and more perfect photographs showed that this was also true of nearly all the stars of the group.

Electra was the mother of Dardanus, the founder of Troy, and the ancestor of Priam and his house. Aghast at the fall of Troy, she fled from her sisters that she might not be obliged to gaze on the destruction of the city so precious in her sight. According to another story she veiled her face so that she could not see the city's fall. Because of these stories respecting her, she has often been regarded as the Lost Pleiad. Ovid called her "Atlantis," personifying the family.

Merope made the mistake of marrying beneath her. Her sisters chose gods for husbands, whereas she selected a mortal, Sisyphus, King of Corinth. She subsequently repented her choice, and hid her tace in shame. On this account she is thought by some to be the Lost Pleiad. Her name signifies mortal, and Allen tells us that the star is enveloped in a faintly extended triangular nebulous haze, visually discovered by Temple in 1859.

Taygeta was the patron goddess of Sparta, since her son Lacedæmon founded that State.

Calæno is said to have been struck by lightning, and consequently is thought by some to be the Lost Pleiad.

Sterope, it is said, married Œnomaus. Their offspring was Hippodaima, a beautiful maiden. The star is a double one, as is Taygeta, and also lays claim to the distinction of being the Lost Pleiad.

Atlas, the father, has his star. Riccioli called the star "Pater Atlas." It represents the mighty man, who, condemned to bear the dome of heaven on his shoulders, was transformed into a mountain. It is a double star and I believe it does not bear claim to be the Lost Pleiad.

Pleione was the mother of the seven sisters, and her star may be the true Lost Pleiad, as the spectroscope reveals evidence of its variable character. It has been suggested that the Lost Pleiad may have been a *nova*, that is, a star which flashed out brilliantly for a time, only to fade away as its fires grew cold.

With the unaided eye seven stars can be seen in this group, although persons possessed of very keen eyesight have been able to count as many as fourteen stars. With a

The Pleiades, Showing Nebula
(Bruce 24-inch Telescope.) Courtesy of Prof. E. C. Pickering

good telescope six hundred stars have been counted, while in a photograph of the cluster taken in 1888 no less than two thousand three hundred and twenty-six stars were revealed.

Of this great galaxy of suns all are drifting across the heavens in the same direction. Two of the stars seem to be hurrying on in advance, like heralds announcing the coming of a host, and six are straggling behind as if wearied by their ceaseless journeying.

The Pleiades are said to be two hundred and fifty light years distant from our system. Our sun removed to this enormous distance would appear as a telescopic star of the tenth magnitude, barely discernible in a three-inch telescope.

Another fact of interest concerning this wonderful star group is that the spectroscope reveals that all these stars are similar in make-up. They all appear to be the product of a common mould, and are in that great class of stars of the Sirian type.

In addition to this the entire group is enshrouded in a nebulous haze, a net that seems to hold its contents fast.

Tennyson well describes the cluster in his line:

> Like a swarm of fire-flies tangled in a silver braid.

Bayard Taylor likened the Pleiades to a swarm of bees upon the mane of Taurus.

Astrologers considered the Pleiades eminent stars, but they denoted accidents to the sight or blindness.

The following list of titles given to this famous star group by the nations of the world ancient and modern attests the fact that of all the stars the Pleiades are the best known and the most celebrated.

NAMES GIVEN TO THE PLEIADES

NAME	SOURCE
The Great Twins	On the Euphrates
A Family Group The Many Little Ones	Babylonians

NAME	SOURCE
Herd of camels	Arabians
The little ones	
A Flame, a Razor	Hindus
Seven sisters of Industry	Chinese
Seven Sisters	The Bible
Seven Stars	
A cluster or heap	
Booths of the maidens	The Rabbis
Flock of Clusterers	Aratos
Rock pigeons flying from Orion	Hesiod
Atlas-born, Seven Virgins	
A Coat of Arms for the merchants	Æschylus
Seven Doves	
Narrow cloudy train of female stars	Manilius
The Rounded Asterism	
Virgin Stars	Virgil
Vergiliæ or Virgins of Spring	The Romans
The Baker's peel or shovel	Gælic
Young Girls	Australians
Wives who shut out their husbands	Hottentots
A season	Polynesians
Tau	
The hoeing stars	South Africans
A company of maidens	Solomon Islanders
Grandfather	Abipones
The Six Stars	S. A. Indians
The Dancers	N. A. Indians
A Sieve	Finns
Mosquito Net	French peasants
The Setting Hen	Russians
Old wives	Poles
Dog baiting a bear	Norse
The Close Pack	Welsh
Starry Seven, old Atlas' children	Keats
Seven Atlantic Sisters	Milton
Hesperides	

NAME	SOURCE
Seven Little Nanny Goats	Sancho Panza
Hen with her chickens Little Dipper A Heap, a troop	Popular names

The Minor Constellations

2 Naos

ζ ARGO NAVIS

ARGO NAVIS

THE MINOR CONSTELLATIONS

Argo Navis	The Ship Argo
Camelopardalis	The Giraffe
Columba Noæ	Noah's Dove
Coma Berenices	Berenice's Hair
Lacerta	The Lizard
Leo Minor	The Lesser Lion
Lynx	The Lynx
Monoceros	The Unicorn
Sagitta	The Arrow
Sextans	The Sextant
Scutum Sobiescanum	Sobieski's Shield
Triangulum	The Triangle
Vulpecula cum Ansere	The Fox with the Goose.

ARGO NAVIS

THE SHIP ARGO

Against the tail of the Great Dog is dragged
Sternward the Argo, with no usual course
But motion contrary.

. . . .

So sternward labours the Jasonian Argo
Obscure in parts and starless, as from prow
To mast, but other portions blaze with light.

FROTHINGHAM'S Aratos.

ARGO can hardly be called a minor constellation, and owes its place under such a heading to the fact that in these latitudes but a very small part of it is visible, so that only a brief reference to it is necessary.

The Ship, or Argo as it is generally called, lies entirely in the southern hemisphere, east of Canis Major and south of the Unicorn and Hydra. Only the few stars representing the stern of the ship can be seen in the latitude of New York City.

The Ship is figured without a prow, one of the best evidences that chance had no part in the invention of the constellation.

According to mythology Argo was built either by Glaucus, Jason, Argos, or Hercules. It was famous as the first craft that ever ventured to sea, and as the one that bore the Argonautic expedition to Colchis on its quest of the Golden Fleece.

To the Egyptians it represented the ark that bore Osiris and Isis over the Deluge.

Sir Isaac Newton fixed the date of the building of this celebrated craft as 936 B.C.

With the Romans it was generally "Argo," or "Navis," and the Arabs called it "a Ship." To the Biblical school it represented Noah's Ark.

The lucida of the constellation, never seen in these latitudes, is the first magnitude star Canopus.

CAMELOPARDALIS

THE GIRAFFE

The Giraffe, described as "a long, faint, and straggling" constellation, first appeared on the star map of Bartsch in 1640. It is also found in the catalogue of Hevelius published in 1690, as Camelopardalus.

Prof. E. C. Pickering tells us that the correct spelling according to the best classical authorities, both Greek and Latin, is Camelopardalis.

Bartsch wrote that the group represented to him the camel that brought Rebecca to Isaac.

The Chinese are said to have located seven asterisms within the borders of this star group.

Photo by Bonfils

Temple of Edfu

According to Argelander the constellation contains eighty-four stars, none brighter than the fourth magnitude.

COLUMBA NOÆ

NOAH'S DOVE

> The surer messenger,
> A Dove sent forth. ..

The Dove first appeared on Royer's star map of 1679, although Allen tells us that it had appeared seventy-six years before on Bayer's plate of Canis Major. It is thought also that Cæsius alluded to it seventeen centuries ago.

The asterism comprises the stars to the south and west of the Greater Dog, and is appropriately situated close to the Ship, which is identified with Noah's Ark.

Alpha Columbæ bears the modern titles "Phaet," "Phaet," and "Phad." The Chinese called this star "the Old Folks."

Lockyer asserts that twelve different Egyptian temples were oriented to this star, notably those at Edfu and Philæ where Phaet was worshipped as far back as 6400 B.C.

Allen considers the star too inconspicuous to warrant such prominence.

Phaet is a 2.5 magnitude star, situated 33° south of ε Orionis, and culminates at 9 P.M., Jan. 26th.

Beta Columbæ is known as "Wezn" or "Wazn," meaning the "weight."

According to Gould, Columba contains seventeen stars.

COMA BERENICES

BERENICE'S HAIR

> Now behold the glittering maze of Berenice's hair.

Eratosthenes was the first to mention this faint yet beautiful cluster, and called it "Ariadne's Hair," but

Tycho Brahe was the first one to catalogue it as a separate constellation in 1602.

Catullus, in his translation of the Greek of Callimachus, thus refers to the location of Coma Berenices:

> Just by the Virgin in the starry sphere,
> The savage Lion and Northern Bear
> Full to the west with sparkling beam I lead,
> And bright Boötes in my course precede.

Allen thinks that the group was known to the Egyptians as "the Many Stars." Other titles for it are "Berenice's Periwig," "Rosa," and "Berenice's Bush," and the figure has been thought to represent the tuft of hair in the Lion's tail, and the sheaf of wheat held by the Virgin.

The Chinese took a great interest in this group, and gave it many fanciful names.

Burritt gives the following brief history of the constellation: "Berenice was of royal descent, and a lady of great beauty, who married Ptolemy Soter or Euergetes, one of the kings of Egypt, her own brother. When he was going on a dangerous expedition against the Assyrians she vowed to dedicate her hair to the goddess of beauty, if he returned in safety. Sometime after the victorious return of her husband the locks, which she had deposited in the temple of Venus, disappeared. The King expressed great regret at the loss, whereupon Conon his astronomer publicly reported that Jupiter had taken away the Queen's locks from the temple and placed them among the stars in this figure."[1]

According to Argelander there are thirty-six stars in this group.

LACERTA

THE LIZARD

This asterism was designed by Hevelius, and comprises the stars between Cygnus and Andromeda.

[1] The early Christians thought that this cluster represented the Scourge of Christ, Absalom's hair, or Samson's Hair.

Photo by Brogi

Berenice
Bronze Bust in National Museum, Naples

The original figure drawn by Hevelius is described as "a strange weasel-built creature with a curly tail."

The Chinese knew the stars in this region of the sky as "the Flying Serpent," and in Royer's chart, published in 1679, these stars formed the star group known as "the Sceptre and Hand of Justice."

Argelander mentions thirty-one stars in this constellation, none brighter than 3.9 magnitude.

Lacerta culminates about the middle of April.

LEO MINOR

THE LESSER LION

> The Smaller Lion now succeeds; a cohort
> Of fifty stars attend his steps.

Hevelius introduced this star group in 1690, forming it from eighteen stars situated between the Greater Lion and Bear. He gave it this title as he said it "partook of the same nature" as the neighbouring figures.

The Chinese included the Greater and Lesser Lion in their great figure of the Dragon.

The zodiacal Crab of the zodiac of Denderah is located in this figure, and this part of the sky was thought to have been sacred to the god Ptah.

Argelander assigned twenty-one stars to this group, none brighter than the fourth magnitude.

LYNX

THE LYNX

The Lynx first made its appearance as a constellation in 1690 on the star map of Hevelius. Originally it was said to contain nineteen stars, which number Burritt has increased to forty-four.

The inventor accounted for the title on the ground that

"it was so inconspicuous a star group that only a lynx-eyed person could discern it."

The Lynx has been known as "the Tiger," and is noted for the number and beauty of its double stars, of which fifty are mentioned in Webb's *Celestial Objects*. The constellation comes to the meridian in February.

MONOCEROS

THE UNICORN

The Unicorn is generally supposed to have been designed by the astronomer Bartsch, but Olbers and Ideler claim that it was invented as early as 1564, and Scaliger is said to have found it on an ancient Persian sphere.

It is situated in the space between Orion, the two Dogs, and the Hydra. The Chinese asterisms, "the Four Great Canals" and "the Outer Kitchen," lay in this region of the sky.

Argelander assigned sixty-six stars to it, and Heis one hundred and twelve.

Monoceros contains many fine star clusters, but no stars brighter than a 3.6 magnitude.

SAGITTA

THE ARROW

There 's further shot another Arrow
But this without a bow. Towards it the Bird
More northward flies.

ARATOS.

This ancient figure is situated in the Milky Way directly north of the Eagle, and has occasionally appeared as held in the Eagle's talons.

"It has been regarded as the traditional weapon that slew the eagle of Jove, or the one shot by Hercules toward the adjacent Stymphalian birds."—ALLEN.

Eratosthenes considered it to be the shaft with which Apollo exterminated the Cyclops, and it has been regarded as the Arrow of Cupid.

In classical times Sagitta was thought to represent the Reed from which arrows were formed.

The Hebrews, Armenians, Persians, and Arabians all knew it as an Arrow.

Cæsius considered it the shaft winged by Joash at Elisha's command, or one of those sent by Jonathan towards David at the stone Ezel.

Schiller thought it represented the spear or the nail of the Crucifixion.

According to Argelander it contains sixteen naked eye stars, none brighter than the fourth magnitude.

It culminates on Sept. 1st.

ζ Sagittæ is a triple star, and "an interesting system," says Allen.

SEXTANS

THE SEXTANT

This is a modern asterism sometimes called "Urania's Sextant," and first appeared on the chart of Hevelius which was published in 1690.

This celestial Sextant is supposed to commemorate the sextant so successfully used by Hevelius in taking stellar measurements at Dantzig from 1658 to 1679.

The astronomer von Rheita imagined that this group represented Saint Veronica's Sacred Handkerchief.

The original figure of the Sextant comprised the twelve unclaimed stars between Leo and Hydra, west of Crater, and it has been generally recognised by astronomers since the date of its invention.

The lucida of the group, a fourth magnitude star, is situated 12° south of Regulus.

According to Argelander the Sextant contains seventeen naked eye stars.

SCUTUM SOBIESCANUM

SOBIESKI'S SHIELD

Hevelius was the first to introduce this figure, which appeared in his star chart of 1690. It is situated in the Milky Way, west of Aquila, between the tail of the Serpent and the head of Sagittarius.

The figure is that of the Coat of Arms of the third John Sobieski, King of Poland, a distinguished warrior.

The group is generally styled "Scutum," and, according to Heis, contains eleven stars, none brighter than the fourth magnitude.

In China these stars comprise an ancient figure known as "the Heavenly Casque."

There are several fine clusters in this region, and it is said that within the boundaries of Scutum, in a space five degrees square, Sir Wm. Herschel estimated that there were 331,000 stars.

TRIANGULUM

THE TRIANGLE

Beneath Andromeda. Three lines compose
The Triangle. On two sides measured equal,
The third side less. It is not difficult
To be discerned. More luminous than many.

FROTHINGHAM'S Aratos.

The Triangle is an asterism of considerable antiquity, and was evidently more noticed by the ancients than by us.

It is situated between Andromeda and Aries, and in the following allusion to it by the poet Manilius there is a reference to its early Greek title, Δελτωτόν, from the likeness the figure bears to the Greek letter Delta (Δ):

Five splendid stars in its unequal frame
Deltoton bears, and from the shape a name.

With the Romans and astronomers of the 17th century it was known as "Deltotum." It was also called "Delta," and associated with Egypt and the Nile, hence its title "the Home of the Nile."

The Triangle has been likened to the Trinity, and the Mitre of St. Peter.

The figure is noted as marking the location of the discovery of the minor planet Ceres by Piazzi, Jan. 1, 1800.

The 3.6 magnitude star Alpha Trianguli bears the title "Caput Trianguli." It culminates at 9 P.M., Dec. 6th.

α and β Trianguli were known as "the Scale Beam."

According to Argelander the group contains fifteen stars.

VULPECULA CUM ANSERE

THE FOX WITH THE GOOSE

This is one of several constellations invented by Hevelius in 1690, and appearing for the first time in his star chart published in that year.

It is situated between the Arrow and the Swan, where the Milky Way divides into two branches.

Hevelius is said to have selected this figure because of its appropriateness to its position, as the fox was a cunning and voracious animal, and was placed near the Eagle and Vulture which are of the same rapacious and greedy nature.

The figure is now generally known as "Vulpecula," and contains a noteworthy object in the "Double-headed Shot" or "Dumb-Bell nebula." The group also marks the radiant point of the "Vulpeculids," a meteor shower appearing from June 13th to July 7th.

According to Argelander the asterism contains thirty-seven stars.

Appendix

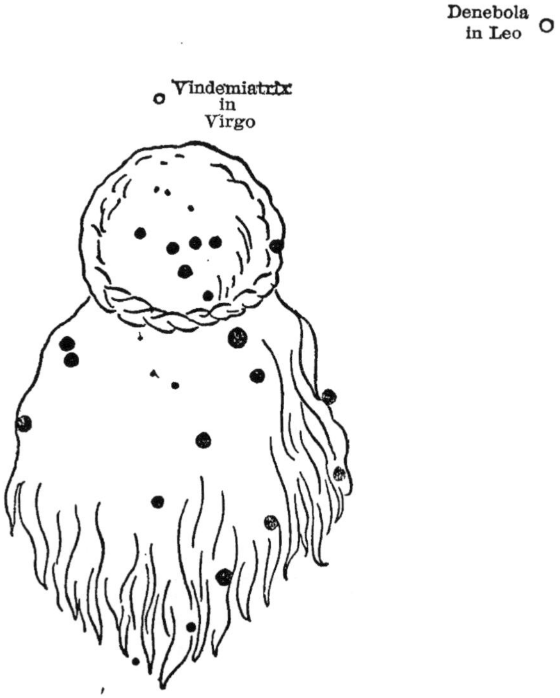

COMA BERENICES

APPENDIX

THE BRIGHTEST STARS VISIBLE IN LATITUDE 40°N.[1]

STAR	MAGNITUDE
The Sun	−25.4
Sirius	−1.58
Vega	0.14
Capella	0.21
Arcturus	0.24
Rigel	0.34
Procyon	0.48
Altair	0.89
Aldebaran	1.06
Pollux	1.21
Spica	1.21
Antares	1.22
Fomalhaut	1.29
Deneb	1.33
Regulus	1.34
Castor	1.58
ε Canis Majoris	1.63
ε Ursæ Majoris	1.68
Bellatrix	1.70
λ Scorpii	1.71
ε Orionis	1.75
β Tauri	1.78
α Persei	1.90
ζ Orionis	1.91
η Ursæ Majoris	1.91
γ Geminorum	1.93

[1] With the exception of the estimate of the sun's magnitude the list is taken from the Harvard Observatory Catalogue.

Other results for the stellar magnitude of the sun are as follows:

Wollaston: −26.6

Bond: −25.8

Zöllner: −26.6

The sun gives us: 10,000,000,000 times the light of Sirius.

LIGHT-GIVING POWER OF THE STARS, SUNLIGHT BEING EQUAL TO UNITY[1]

SIRIAN STARS	
Procyon	25
Altair	25
Sirius	40
Regulus	110
Vega	2050

SOLAR STARS	
Aldebaran	70
Pollux	170
Polaris	190
Capella	220
Arcturus	6200

The total light of the stars is estimated as equal to $\frac{1}{80}$ of that of the full moon.

NEAREST LUCID STARS IN THE NORTHERN HEMISPHERE

Star Name	Magnitude	Distance in Light Years according to Todd	Russell	Gore	Yale Univ.	Young	Newcomb
Sirius	−1.6	8.5	8.6	9		8.6	8
τ Ceti	3.6		9.7			10.2	10
Procyon	0.5	12	10	10	9.8	10.9	10
61 Cygni	5.6	7.2			11.1	8	7.3
Altair	0.8	16		14	14.1	13.6	14
Vega	0.1	27		40		21.7	29
Aldebaran	1.0	32		32	28	29.6	29
Capella	0.2	32		40	34		36
Polaris	2.1	47		46		44	54
Arcturus	0.2	160		160			108
β Cassiopeiæ	2.4			32			21

Of the fainter stars in the northern hemisphere the 7.4 magnitude star Lalande 21185 is probably the nearest star to the earth. The average distance as estimated by different authorities is 7.5 light years.

The distance of the first magnitude star α Centauri in the southern hemisphere, probably the nearest star to the earth, is given by all authorities as 4.3 light years. This distance is better realised if we adopt Prof. Young's comparison: If the distance from the earth to the sun were 215 ft. the distance from the earth to α Centauri would be 8000 miles.

On the scale measured at Yale University the mean distance of stars of the first magnitude is 36.5 light years, second magnitude stars 58 light years, and those of the third magnitude 92 light years.

[1] From calculations made by Maunder.

NUMBER OF THE STARS[1]

First Magnitude	20	
Second "	65	
Third "	200	
Fourth "	500	
Fifth "	1400	
Sixth "	5000	
Seventh "	20000	
Eighth "	68000	
Ninth "	240000	
Tenth "	720000	1,055,185

The lucid, or naked-eye, stars comprise the first six magnitudes.

A 5″ telescope reveals stars down to the 12th magnitude, and Prof. Ritchey of the Mt. Wilson Observatory using the new 60″ reflector has photographed by four-hour exposures stars probably as faint as the 20th or 21st magnitude. It has been estimated that the total number of stars within our ken photographically speaking is possibly 125 million.

Oldest Stars (Red)	Next in Order (Yellow)	Youngest Stars (White)
Antares Aldebaran Betelgeuse	Our Sun Capella Pollux Arcturus	Sirius Deneb Procyon Spica Altair Regulus

PERIODIC COMETS

Name	Last Perihelion	Period in Years	Next Return
Encke	Sept. 15, 1901	3.3	
Brorsen	Feb. 25, 1890	5.45	1911
Tempel Swift	June 5, 1897	5.54	1913
De Vico Swift	Apr. 27, 1901	6.4	1914
Tempel II	Oct. 4, 1898	6.5	1911
Finlay	Feb. 17, 1900	6.5	1913
Wolf	July 5, 1898	6.8	1912

[1] From Todd's *Astronomy*.

PERIODIC COMETS

Name	Last Perihelion	Period in Years	Next Return
Holmes	Apr. 29, 1899	6.8	1913
Faye	Jan. 23, 1881	7.5	1911
Tuttle	May 5, 1899	13.6	1913
Pons Brooks	Jan. 26, 1884	71.5	
Olbers	Oct. 19, 1887	72.6	
Halley	May 17, 1910	76	

PROPER MOTION OF THE STARS
(The angular motion across the line of sight.)

Star Name	Speed in Miles per Second	
	Pritchard	Young
β Cassiopeiæ	10	
α "	2	
61 Cygni	35	37
Polaris	2.5	1.8
α Arietis	8	
α Persei	1	
	Elkin	
Aldebaran	4	5.1
Capella	11	
Sirius	9	10.1
Procyon	13	12.2
Pollux	27	
Regulus	8	
Vega	31	7.1
Altair	9	8.0
	Miss Clerke	Newcomb
Arcturus	375	200 to 300

SECCHI'S SPECTROSCOPIC STAR TYPES

	Characteristics	Star Names
Type I. Sirian Stars (blue or white)	Broad, intense, dark hydrogen lines	Sirius Vega Altair and perhaps more than half of all the stars

SECCHI'S SPECTROSCOPIC STAR TYPES

	Characteristics	Star Names
Type II. Solar Stars (yellowish like sun)	Fine, dark, metallic lines	Capella Arcturus
Type III. Orange and reddish stars	Many dark bands	α Herculis Mira Antares A majority of the variable stars
Type IV. Blood red in tint	Dark bands or flutings, the reverse of Type III. as to shading	About 50 stars of this type
Type V.	Bright lines	Number about 70 Situated near the middle of the Galaxy

SCINTILLATION

SECCHI'S TYPE	MEAN SCINTILLATION
I	87
II	79
III	59

Scintillation is most pronounced in January and February, and magnetic storms and violent scintillations are absolutely coincident in point of time.

STARS APPROACHING THE EARTH

Star Name	Speed in Miles per Second			
	Potsdam	Todd[1]	Greenwich	Vogel
α Arietis		11.7		9.2
γ Leonis		25.1		
Spica	14	10.6	17	9.2

[1] From Todd's *Astronomy*.

STARS APPROACHING THE EARTH

Star Names	Speed in Miles per Second			
	Potsdam	Todd[1]	Greenwich	Vogel
Altair		23.9	27	23.7
Polaris	16.3		16	
Algol	2.3		2	
Arcturus			45	4.6
Vega			34	9.7
Deneb			36	5.1
Pollux[2]			33	
Sirius				9.7
Procyon	7			5.5
Castor				18.4

STARS RECEDING FROM THE EARTH

Star Name	Speed in Miles per Second			
	Potsdam	Todd[1]	Vogel	Greenwich
Aldebaran	30	31.1	30.1	31
Rigel	39	13.6	10.1	18
Betelgeuse		17.6		28
α Coronæ		20.3		
Capella	17		15.2	23
ε Orionis	34		35	15

FAMOUS TEMPORARY STARS

Date	Appearing in the Constellation
134 B.C.	Scorpio The Star of Hipparchus
123 A.D.	Ophiuchus

[1] From Todd's *Astronomy*.

[2] According to Allen Pollux is receding from the earth at the rate of 1 mile per second.

FAMOUS TEMPORARY STARS

Date	Appearing in the Constellation
386	Sagittarius
389	Aquila near Altair
393	Scorpio
1012	Aries
1203	Scorpio
1230	Ophiuchus
1572	Cassiopeia Tycho's Star
1604	Ophiuchus Kepler's Star
1670	Vulpecula
1848	Ophiuchus
1860	Scorpio
1866	Corona Borealis
1876	Cygnus
1885	Andromeda
1891–92	Auriga

FAMOUS VARIABLE STARS

Star Name	Period in Days	Range in Magnitude	Type
o Ceti	331	1.7 to 9.5	Mira
R. Leonis	313	5.2 to 10.0	
β Persei	$2\frac{7}{8}$	2.3 to 3.5	Algol
ζ Geminorum	$10\frac{1}{7}$	3.7 to 4.5	
δ Libræ	$2\frac{1}{3}$	5 to 6.2	
χ Sagittarii	7	4 to 6	
β Lyræ	12.9	3.4 to 4.9	
α Herculis	90±	3.1 to 3.9	Irregular

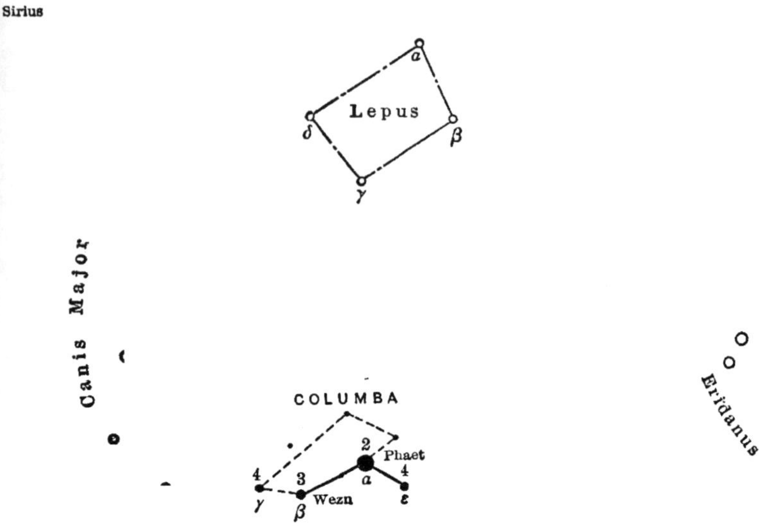
Sirius
Lepus
α
β
γ
δ
Canis Major
Eridanus
COLUMBA
2
Phaet
4
3
4
Wezn
α
β
γ
ε

COLUMBA

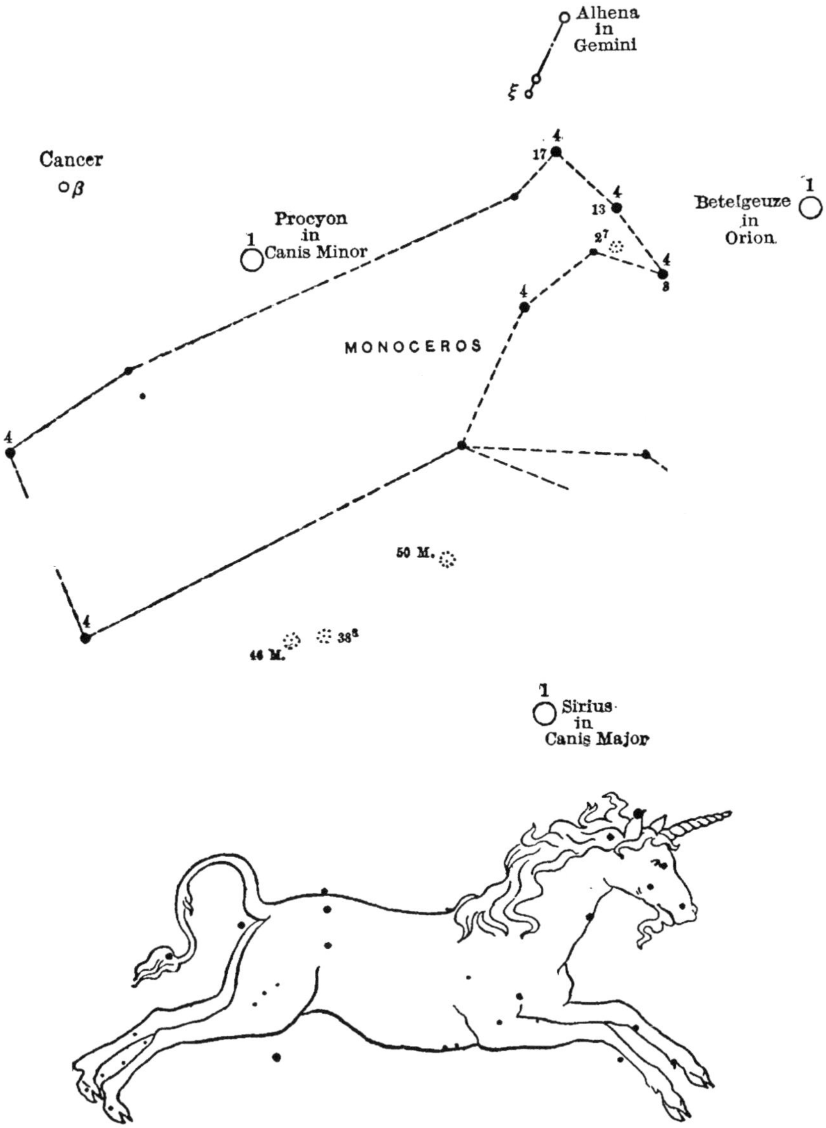
Alhena in Gemini
ξ
4
17
Cancer
β
Procyon in Canis Minor
1
4
13
Betelgeuze in Orion
1
2⁷
4
8
4
MONOCEROS
4
4
50 M.
4
46 M.
38⁸
1
Sirius in Canis Major

MONOCEROS

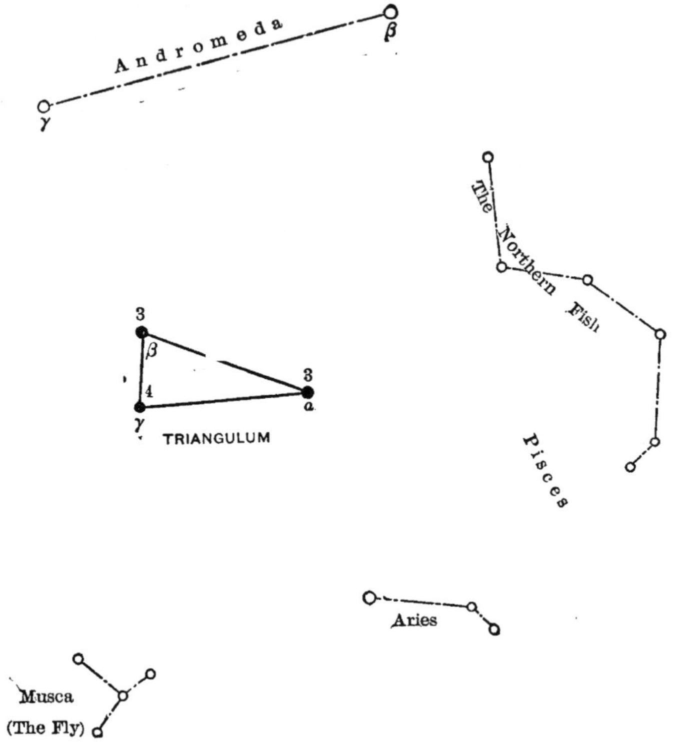
Andromeda
β
γ
The Northern Fish
Pisces
3
β
3
4
γ
α
TRIANGULUM
Aries
Musca
(The Fly)

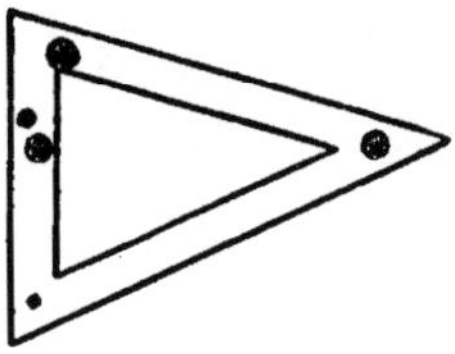

TRIANGULUM

INDEX TO CONSTELLATIONS

Spotted something that isn't quite right?

Let us know and we'll fix it for future reprints. And as a thank you for your diligence, we'll send you a free Sothis book of your choice.

publisher@sothis.press

This page intentionally left blank

Made in United States
Orlando, FL
04 November 2026

90204526R00337